20 YEARS AT THE TOP
A Generation of Black Engineers of the Year

By the Staff of Career Communications Group Inc.

© Copyright 2006 by Career Communications Group

All rights reserved. No part of this book may be reproduced in any form or by any electronic or mechanical means, including information storage and retrieval systems, without permission in writing from the publisher, except by a reviewer who may quote brief passages in a review.

Published by Career Communications Group Inc. Inquiries should be sent to Career Communications Group Inc., 729 E. Pratt St., 5th Floor, Baltimore, Maryland, 21202

ISBN: 1-4116-7685-8

Second Edition

Contents

The First Decade

Foreword 7

Chapter 1
John Brooks Slaughter - 1987 Black Engineer of the Year 11

Chapter 2
Erroll B. Davis Jr. - 1988 Black Engineer of the Year 15

Chapter 3
Captain Donnie Cochran - 1989 Black Engineer of the Year 19

Chapter 4
Arlington W. Carter - 1990 Black Engineer of the Year 25

Chapter 5
Guion S. Bluford Jr. - 1991 Black Engineer of the Year 31

Chapter 6
Arnold F. Stancell - 1992 Black Engineer of the Year 37

Chapter 7
James W. Mitchell - 1993 Black Engineer of the Year 41

Chapter 8
William R. Wiley - 1994 Black Engineer of the Year 45

Chapter 9
Walt W. Braithwaite - 1995 Black Engineer of the Year 49

Chapter 10
Albert J. Edmonds - 1996 Black Engineer of the Year 55

The Second Decade

Chapter 11
Arthur E. Johnson - 1997 Black Engineer of the Year 61

Chapter 12
Joe N. Ballard - 1998 Black Engineer of the Year. 65

Chapter 13
Paul L. Caldwell - 1999 Black Engineer of the Year. 71

Chapter 14
Mark E. Dean - 2000 Black Engineer of the Year 101

Chapter 15
Shirley Ann Jackson - 2001 Black Engineer of the Year 107

Chapter 16
Rodney O'Neal - 2002 Black Engineer of the Year 115

Chapter 17
Lydia W. Thomas - 2003 Black Engineer of the Year. 121

Chapter 18
Anthony R. James - 2004 Black Engineer of the Year 125

Chapter 19
William D. Smith - 2005 Black Engineer of the Year. 131

Chapter 20
Linda Gooden - 2006 Black Engineer of the Year 137

Foreward

Sometimes it is difficult to be inspired by the past. The great men and women in African American history too often seem devoid of personality. We read of their triumphs and breakthroughs, but we cannot imagine what they felt at the time. Was the discovery of the gamma electric cell as exciting for Henry T. Simpson in 1971 as it seems more than twenty years later? Could Thomas J. Martian have imagined the magnitude of his invention of the fire extinguisher? It is hard even to imagine life without fire extinguishers today. And while driving, how many of us remember Richard B. Spikes, the man who invented the automatic gearshift? We find it easier to make role models of those people we can touch, and listen to, and question. In the midst of today's negativity, African Americans everywhere are accomplishing great things.

These people serve as the role models for youth in their communities and for anyone else who has the opportunity to meet them. We can listen to their struggles, feel their disappointments, and share in their successes. Some of today's giants are profiled here, receiving the honor and recognition they truly deserve. Some slammed into the technical and corporate establishment and crashed through its barriers, while others patiently removed the stones laid to obstruct their progress. All of them are innovators who have made deep impressions on their peers and colleagues because they took risks and overcame difficult challenges. With intelligence, enthusiasm, and most of all effort, these men and women have succeeded in their fields. All of them are winners.

—Yvonne R. Keith, former associate editor, *USBE&IT* magazine

Part I:
The First Decade

Chapter 1

John Brooks Slaughter
1987 Black Engineer of the Year

Dr. Slaughter, a product of schools still segregated during the 1940s, had met stunned laughter when he told his high school counselors he wanted to be an engineer. Neither he, nor they, had ever seen a Black engineer, and they tried to steer him into vocational courses. But the young John Slaughter would not only become an engineer, but also go on to a distinguished career as an educator.

Beginning as an electronics engineer at General Dynamics—a company that made everything from supersonic airplanes to submarines—the young John Slaughter moved steadily up the ranks, crossing the lines from industry, to government, to academia, and back again. Bonnie Winston reported in a 1987 US Black Engineer & Information Technology magazine story that when Slaughter joined the Naval Ocean Systems Center in San Diego, Calif., in 1960, he was refused a supervisory position.

Two years later, Slaughter was running a department with 250 researchers, including the man who was hired for the supervisory job Slaughter had been denied. Slaughter simultaneously stood duty as director of the Applied Physics Laboratory at the University of Washington. His team developed some of the early theories for computer control systems in naval weapons. Another Slaughter-led initiative delved into underwater acoustics research and applications that had critical implications in the decades-long confrontation between U.S. forces and submarines from the Communist-led Soviet Union.

Down the Pacific Coast, at the University of California, he completed studies for his master's and doctorate of philosophy in 1971. Six years later, Dr. Slaughter, then one of only three Blacks elected to the prestigious, 1,100-member National Academy of Engineering, moved east to Washington, D.C., as assistant director of the National Science Foundation. In 1979, President Carter named him director.

When John Slaughter stepped up to the leadership of the federal agency driving American science—like his older contemporary, Lincoln Hawkins, the first Black member to be elected to the National Academy of Engineering—Dr. Slaughter became a mentor and driving force for promotion of Blacks and other minorities in the research establishment. The legion of people he helped during his climb to the top of American science cheered when *US Black Engineer & Information Technology* magazine named Dr. Slaughter the first Black Engineer of the Year Award (BEYA) winner in 1987.

His BEYA citation included many of his achievements. As fellow of the American Association for the Advancement of Science, he served on the Committee on Minorities in Engineering for the National Academy of Engineering. In 1977, he served as editor of the *International Journal of Computers and Electrical Engineering*, and he still chairs the Action Forum on Engineering Workforce Diversity. He's also a fellow of the American Academy of Arts and Sciences and the Institute of Electrical and Electronics Engineers, and in 1993 the American Society for Engineering Education inducted him into its Hall of Fame.

After a long, distinguished tenure at Occidental College, Dr. Slaughter moved back East in 2003 to take over as president and CEO of the National Action Council on Minorities in Engineering (NACME), founded 32 years ago out of a National Academy of Engineering inquiry on the low status and small numbers of Blacks, Latinos, and Native Americans in the engineering profession. A 2003 commemorative banquet at New York's Waldorf Astoria Hotel raised new money for NACME, which has spent more than $100 million to boost some 18,000 new minority entrants into the technology work force, paying their college costs and sponsoring internships to prepare them, dramatically changing the face of engineering.

Today, with American industry facing an aging work force, shortfalls in federal funding for research in physical, mathematical sciences, and engineering, and a decline in interest in engineering careers, NACME's role in providing scholarship aid to minority degree candidates is critical. In the 2004–2005 academic year, NACME recorded a 40 percent increase in scholarships to 733 students, 36 percent of whom are women. Latinos increased 9 percentage points to 44 percent of the total, while Blacks represent 49 percent, and Native Americans, 3 percent.

But NACME is more than a scholarship agency. It is a producer of research analyses on the state of the technology work force and minorities in industrial America, which serves as a sounding board and advocate for initiatives to increase the number of minorities in the workgroups and career tracks responsible for some 50 percent of the country's economic growth. Its new Research Council, composed of leaders in engineering, is designed to maintain and strengthen NACME's role as a provider of information on minority performance in math and science at the secondary school level, and on their success in university level engineering programs.

THE FOLLOWING ARE EXCERPTS FROM A 2000 US BLACK ENGINEER & INFORMATION TECHNOLOGY *MAGAZINE INTERVIEW WITH DR. JOHN SLAUGHTER, CONDUCTED BY TYRONE TABORN AND GARLAND THOMPSON.*

I was privileged to receive the first Black Engineer of the Year Award. I recall the evening of the award presentation vividly. I remember thinking about how my parents had sacrificed and worked to make it possible for me to pursue my dream of becoming an engineer. I wished that they could have been there to experience the event, because it was because of them and because of the unwavering support of my wife, Bernice, and my children, John Jr. and Jacqueline, that I was standing there to receive the award. On that occasion, none of us present could have envisioned how significant and widely recognized this annual program would become.

I felt equally privileged to participate in this year's recognition of Dr. Mark Dean of IBM as the Black Engineer of the Year 2000. Mark's contributions to the fields of computer science and information technology are truly exemplary. As a member of the team that created the IBM PC, the machine that has transformed the way in which we live, and a leader in the development of technology for a network-connected digital world, Dr. Dean was singularly deserving of this award.

To be sure, the situation today for engineers of color is much improved over what it was when I began my engineering career. Then, the term "Black engineer" was nearly an oxymoron.

In 1956, when I graduated from college with a shiny new baccalaureate in electrical engineering, the proportion of African Americans in the engineering work force in this country was less than one-half of a percent. It is no joke when I say that I was the first Black engineer I ever met.

The launching of Sputnik and the subsequent "Space Race," as well as an increasing public awareness of the huge loss in human potential that was occurring as a result of discriminatory hiring practices in industry, brought about slow but inexorable change. Today, the percentage of Black engineers is three percent—an improvement, but still far less than the proportion of African Americans in our nation's population.

The reasons for this are many, but none is as significant as the dearth of Black young men and women who are encouraged and prepared in our public schools to enroll in the mathematics and science courses that are prerequisites for engineering study. Many schools, particularly those in the inner cities of America, are ill-equipped and have unprepared and unqualified math and science teachers. These schools simply cannot provide the instruction needed by children. We can no longer sit by and let this happen to our young people. We have tolerated it for too long.

Recently, the National Academy of Engineering launched a new effort to involve industry, government, and higher education in addressing the need for increased diversity in the engineering work force. The creation of this new "Action Forum" is another indication of the increasing recognition of the importance of strengthening the profession by including more of those who historically have been under-represented.

Throughout history, beginning with slavery, Black Americans have played an important role in the building of this country. Agriculture, transportation, and construction, to mention but a few enterprises, all have benefited, often unfairly, from the efforts of Black men and women. Today, a new set of opportunities awaits those who are prepared. It is the task of this generation to help prepare those who will build America throughout this newly dawned century. I can think of no more important or exciting assignment.

Chapter 2

Erroll Davis Jr.
1988 Black Engineer of the Year

Erroll B. Davis Jr., a B.S.-holder in electrical engineering from Carnegie Mellon University in his native Pittsburgh, won an IBM Fellowship to complete his M.B.A. studies at the University of Chicago in 1967 before launching a career that made him the highest ranking Black American executive in the utility business. Davis, a Reserve Officer Training Corps cadet at Carnegie Mellon, served in the Army's Tank Automotive Command at a maintenance facility in Warren, Mich., from 1967 to 1969. The Vietnam conflict was raging, and many in Davis' generation were heading off to fight in Southeast Asia, but Davis jokes that the Army sent him to Michigan because of his poor eyesight. Davis spent his time learning how to manage a multimillion-dollar technology organization and digging into what it took to keep his hardware running at optimal levels. He didn't go far from his Michigan posting when he mustered out. After fulfilling his ROTC active duty obligations, Davis joined Ford Motor Co., an outfit whose reputation for opening up career opportunities for minorities goes all the way back to Henry Ford's leadership during the 1930s.

He worked for four years in the Finance Department of Ford Motor Company. Then, he left for a slightly longer stint at Xerox Corporation, another integrated manufacturer. Again, he served on the finance staff, moving up as he learned what it takes to keep Xerox humming.

Davis signed on at Wisconsin Power & Light (WP&L) in 1978 as vice president of finance, joining his contemporary Cordell Reed at Chicago's Commonwealth Edison in a tiny class of Black utility executives. Reed, a South Side public housing alumnus who was only the third Black American to complete the mechanical engineering major at the University of Illinois-Urbana, had in 1975 become a department head and public spokesman for nuclear power. Now retired, Reed is a widely respected member of the National Academy of Engineering.

Just up the road in Madison, Wis., Erroll Davis was racing ahead. He joined WP&L the same year a young Anthony James (2004 Black Engineer of the Year) signed on as a safety and health supervisor at Georgia Power, adding one more to the small but growing corps of Black utility managers.

Davis began a double assault, making his way up the leadership ladder in community activities as he charted a careful charge up the corporate ranks. Upon arriving in Madison, he volunteered in local United Way activities and the Madison Art Center, and he and his wife established a college scholarship fund for Black students.

By 1987, when Davis rose from executive vice president to president of WP&L—the nation's only Black leader of a publicly held company—he had reached the rank of chairman of the United Way of Dane County, Wis. He also was a director of the Wisconsin Manufacturers and Commerce organization, a member of the Madison Police and Fire Commission, a director of the Madison Capital Corp. and the Madison Development Corp., and past president of the Madison Urban League Board of Directors.

Black Enterprise magazine took notice: Davis was now the highest ranking Black in any business field. Asked about his new high status by a local Madison news reporter, Davis said he was proud to be Black, but he didn't dwell on race. He had much to do and learn to be successful. Or, as he put it, "I could also be the first Black president to be fired."

Davis, now running all parts of the utility's operations, stepped into the top job at a time when then-CEO James R. Underkofler and the board of directors were pushing restructuring of the corporation. Their plan, now complete, created a holding company for utility operations and subsidiary corporations, to pursue developing opportunities in such businesses as landfill development and communications.

They had to move fast. WPL Holdings, as it became known, faced growing competition from other gas and electric suppliers as energy deregulation opened doors to non-utility generators and the selling of excess power from private corporate systems into the public net. Large customers were beginning to buy natural gas directly from its producers, using utilities such as WPL only to transport the gas through their pipelines.

In other words, WPL, a company employing 2,600 workers supplying power and natural gas to customers spread over a 16,000-mile part of southern Wisconsin, faced large challenges. Revenues had slipped from $170.2 million in the first quarter of 1986 to $157.4 million, only partially because of an unusually warm winter.

Davis, a key player in the transition to a holding company and then in the merger with several other utilities to become the main electric power and gas supplier for southern Wisconsin and parts of Iowa, Illinois, and Minnesota, was ready for his move up to the last rung of the corporate ladder.

Now, Davis is chairman and chief executive of Alliant Energy, a Fortune 1000 company with 8,500 employees across the country and abroad. Alliant, traded on the New York Stock Exchange under the ticker symbol LNT, had operating revenues of $3.1 billion in 2003, and total assets of more than $7.7 billion. The company serves more than 1.4 million customers over a 54,000-square-mile territory that contains nearly 10,000 miles of electric transmission lines and 8,000 miles of natural gas mains. Alliant operates fossil fuel, nuclear, and renewable generating facilities across the upper Midwest, generating more than 31 million megawatt hours of electric power a year.

Davis is also a member of the Board of Trustees of Carnegie-Mellon University and a director of B.P. Amoco, PPG Industries, and Union Pacific.

In June of 2004, Davis was chosen to be one of only four independent directors on the board of the U.S. Olympic Committee (USOC). He is serving a six-year term on the 11-member board chaired by former Los Angeles Olympic Organizing Committee head and Baseball Commissioner Peter Ueberroth. The USOC says the naming of this board, and its reduction from 125 members to 11, represented "the final step in the most sweeping governance transition in the history of the U.S. Olympic Committee."

Other steps included naming of an Olympic Assembly, dividing responsibilities between management and the board of directors, and redefining the USOC's mission to emphasize providing support for America's Olympic and Paralympic athletes.

In a nutshell, Erroll Davis—named by *US Black Engineer & Information Technology* magazine and *blackmoney.com* as one of the "50 Most Important African Americans in Technology in 2003" and as one of the "50 Top Blacks in Technology" in 2004, again by *USBE&IT* is a man at the top of his game.

Chapter 3

Lt. Comdr. Donnie Cochran
1989 Black Engineer of the Year

Donnie Cochran always wanted to fly. Growing up on a family farm near Pelham, in Georgia's southwestern section, young Donnie, the fifth of 12 children, paused while working in the fields to watch Navy jets fly over.

"I started thinking about flying when I was about 12 years old," he told *US Black Engineer & Information Technology* magazine in a 1987 interview. "Out working in the fields, I had the opportunity to see Navy jets flying over, in what I later figured out was low-level navigation training. And as I'm out there, working in the heat of the day, I see those jets screaming by, and I wondered, 'Now which is better, me down here or those jets up there? Which is more exciting?' It didn't take me long to realize that if I had an opportunity to pursue flying, that was what I was going to do."

An older sibling came home to Pelham wearing Naval Reserve Officer Training Corps whites in the summer of 1972, just after Cochran finished high school. Three older brothers had gone on to historically Black Savannah State College and majored in engineering, and when young Donnie saw his oldest brother "wearing this real sharp white uniform, I thought, 'this is pretty neat,' and then he was going traveling, something I hadn't done. I hadn't traveled out of the state of Georgia."

Cochran majored in civil engineering technology, and might well have continued the tradition begun by pioneers such as Archibald Alexander, an internationally known Black architect and bridge-builder responsible for the reflecting pool at the Jefferson Memorial, Washington's Whitehurst Freeway, and the Alabama airfield where the Tuskegee Airmen got their wings.

But flying was in Donnie Cochran's blood. Coach Roscoe Draper would understand that drive implicitly. Coach Draper, a colleague of the now-legendary Chappie James and Cap Anderson, learned to fly in the Army's Civilian Pilot program begun before World War II, but like other Blacks, was not allowed to join the Air Corps. Draper, undeterred, qualified for his instrument and instructor ratings, and earned his nickname of "Coach" teaching the Tuskegee Airmen to fly. Coach Draper now lives in Philadelphia and is an active member of the Black Pilots of America chapter that bears his name. More than 60 years after his famous students wrote their names into history, he continues to fly and works occasionally as a licensed FAA aircraft inspector.

Donnie Cochran enrolled in Savannah State's Naval ROTC, too, and joined the FLIP program, which put him into the cockpit of a Cessna 152, following a similar path as that of Draper and the Air Force's Chappie James. By the time Cochran graduated, he had even made a "hop" in a TA-4 tactical jet trainer, and completed 30 hours of flying Cessnas. Using his degree and NROTC credits, Cochran skipped the rigorous Aviation Officer Candidate School and graduated into a commission.

After that came flight school at Pensacola, the closest Naval Air Station from which the planes Cochran admired as a youth could have come. It didn't hurt that he'd majored in civil engineering. Military pilot school provides the equivalent of master's degree training in the dynamics of flight, structural properties of airframes, physics of gas turbine engines, communications, computers and aviation electronics, and airborne weapons systems.

After basic and advanced jet training at Kingsville, Texas, Cochran went to sea, flying RF-8 Crusader photographic reconnaissance aircraft off the decks of the nuclear-powered carrier Nimitz in the Mediterranean Sea. Photo reconnaissance is critical at such times, when fleet commanders and their civilian leaders at the Pentagon need up-to-date information on the deployments of potential enemy forces and the possible sites of American military incursions. Cochran and his reconnaissance comrades were in high demand.

Cochran shifted to the F-14 Tomcat, the Navy's premier air superiority fighter, for 38 months of sea duty aboard the USS Enterprise, based just north of San Diego, Calif., at Miramar Naval Air

Station, memorialized in *Top Gun* as "Fighter Town, U.S.A." The conditions under which the Blue Angels fly are even more hazardous. A summer 1987 *US Black Engineer & Information Technology* magazine story by Grady Wells noted that the Angels—the world's most famous precision aerobatics team—have taken the art of close-quarters formation flying to unprecedented levels: "Sixty-three times a year, flying over 37 different locations, the Blue Angels push the performance envelope in their distinctively painted F/A-18 jets. Swooping over onlookers' heads in their patented diamond formation, smoke trails streaming behind, the Angels then perform echelon rolls, double farvels, diamond rolls, clean and dirty loops, Cuban eights, four-point hesitation rolls, opposing blivots, their trademark fleur de lis, and the complex delta vertical break, which ends with all six planes crossing simultaneously in front of the audience at speeds exceeding 500 nautical miles per hour, at 'minimum separation.'"

The dangers of their work became all too apparent on July 13, 1985, when two Blue Angels A-4 Sky Hawks collided over Niagara Falls Air Force Base. Lieutenant Commander Mike Gershon, of Pensacola, Fla., was killed, and Lieutenant Andy Caputi ejected and parachuted to safety on the grounds.

Lt. Commander Cochran and two other pilots were chosen to replace the two men. On Oct. 4, 1985, Cochran became the first Black man to fly in the Blue Angels formation in the team's 40-year existence. Cochran had flown more than 2,000 hours in jet fighters and completed 469 carrier landings. He was 31 years old.

On July 4, 1986, Cochran and his teammates flew Sky Hawks in a ceremonial celebration to salute the restoration of the Statue of Liberty. The young officer, finding himself doubly the center of attraction because of his historic posting, told reporters, "What I am doing is not just a job, it's an opportunity. I would like to show young people the roads that are open to them in America. Nobody said, 'here, Donnie, apply for the team, and they will give it to you.' You have to earn it."

Interestingly, when *US Black Engineer & Information Technology* magazine interviewed him a year later, Cochran didn't seem to think his being chosen was all that unusual:

"They were looking for people who could do the job. . . . I think about 70 percent of all Naval tactical aviators who fly off ships could be trained to do what we do. That says a lot about the way the Navy trains its pilots."

After his two-year tour with the Blue Angels, moving up to the newer F/A-18 Hornet, Cochran went back to Miramar Naval Air Station and joined the "Bounty Hunters" of Fighter Squadron VF-2, deployed aboard the carrier USS Ranger, Admiral Davis' former command. Then he was selected to attend the Air War College in Montgomery, Ala., and also earned a master's degree in human resource management from Troy State University.

Duty called again, and in March 1992, Cochran reported to Fighter Squadron VF-1 as executive officer. The next year, he took command, leading the squadron until it was disestablished, then moved on to command the "Sun Downers" VF-111.

History caught up with Cochran again when the explosive news reports of wild behavior at the Navy fliers' annual Las Vegas "Tailhook" reunion resulted in the grounding of the Blue Angels' leader, Commander Robert E. Stumpf. Comdr. Cochran, 41, was called back to take over as leader in 1994.

"I look at [being the first Black man to lead the Blue Angels] as an opportunity to be the boss," he said then. "Not Black and white, but an opportunity to command a very special organization," consisting of not only the six formation flyers, but also the ground support team and the Marine aviators flying the C-130 transport that accompanies the group and is an integral part of every Air Show performance.

In June 1996, Cochran decided to leave the Blue Angels. Painfully aware that 22 Blue Angels pilots died in training or in air shows before he joined the team, Cochran had stood down the unit twice to strengthen safety procedures.

"I can hold my head high," he said. "I have not crashed any airplanes, none of my pilots have crashed an airplane, none of my pilots have been hurt."

Captain Cochran had spent more than 4,350 hours in seven different types of aircraft, had completed 570 carrier landings, and had been awarded the Meritorious Service Medal, the Air Medal, and the Navy Commendation Medal, among numerous other awards.

Chapter 4

Arlington W. Carter
1990 Black Engineer of the Year

Born in the windy city of Chicago, Arlington W. Carter was educated at a technical high school. He showed what he called "a reasonable interest toward science, math, physics, and chemistry," and went on to junior college. Then the drums of war rolling across the Korean Peninsula beckoned. In the Air Force, Carter spent four years as a weather forecaster.

After discharge, Carter decided to be an engineer. Unsure which field to choose, he picked electrical engineering "after reading electrical engineers made more bucks." In addition to making money, electrical engineers working in 1961, when Carter finished at Illinois Institute of Technology, were adapting their new solid-state electronics gear to swing knockout blows at the old ways of doing business. And nowhere was that shift so apparent as in aerospace, where Carter wound up.

Carter joined Boeing because he liked the high-tech atmosphere, and he quickly got his hands into activities that had nothing to do with his chosen field but everything to do with how life would be lived in the world he was helping to fashion. Starting out as a Minuteman strategic missile specialist and moving on to the Advanced Surface to Air Missile System, the Safeguard/Sentinel program, and the Strategic Silo Upgrade program, he reached out into community activities inside and outside the corporation.

Interviewed by *US Black Engineer & Information Technology* magazine in 1989, Carter said young engineers looking to move up should jump right into community service:

"I believe one of the best places to learn how to manage is through volunteer efforts, such as charities, church activities, and politics. If you can learn to manage volunteers, you certainly learn the techniques for managing within your company. You'll learn techniques of how to coach, appraise, model, and inspire. These attributes are critical for moving up, for showing you have the maturity and poise. There are some things you don't learn in school."

Clearly, Arlington Carter learned those things well. During the early 1970s, he spent two years as a loaned executive for the City of Seattle. There, he established and managed Seattle Housing Development, a nonprofit corporation that built and refurbished housing for low- to moderate-income residents. He had a $2.5-million annual budget and a staff of 36.

Occasionally, he found that the business interests of Seattle Housing Development were diametrically opposed to those of Boeing, his employers. To complicate matters more, Carter worked for a Democratic mayor under a Republican governor, and they often disagreed on policy.

"I had to learn to walk a delicate line in a situation that was more political than corporate," he told *USBE&IT*. "But it was rewarding because it helped me to learn about politics."

It also gave him good management experience. His staff included individuals who frequently had less education and fewer job skills than his corporate co-workers. Worse, when he did get someone properly trained, that person took off for higher pay and benefits in the corporate world. Finally, Carter found that contractual relationships and lines of authority, so clear and compelling in a corporate setting, were often more diffused.

"It was important to be more responsive to community needs, to citizens, and various communities. You have to inform what you're trying to do and try not to offend," he said.

In the end, it paid off. In 1978, Carter won his first in-house executive management position as general manager of the Seattle Service Division, a then-new branch with more than 10,000 workers. It not only made him the first Black to serve on Boeing's Executive Management Team, but also gave him significantly broader responsibilities.

Earlier, Carter had worked strictly on technology projects. In addition to his work on missile systems, he also managed a $25-million R&D contract to test high-speed locomotives for the

Federal Railway Administration. "In the overall corporate picture, it was not a big job," he said, "but it was significant for me because I controlled the engineering, purchasing, contract—the whole works."

Putting that together with his experiences managing disparate skill sets and dealing with multiple constituencies through Seattle Housing Development, Carter knew what to do in his new role.

Other Boeing executives had struggled to make the human resources department's processes work better to provide more direct support to the staffing of individual projects. Under Carter, the department simultaneously staffed two major new airplane programs, the Boeing 757 and 767, which are mainstays in today's air transport industry. Carter's team hired more than 20,000 people with skills in engineering, computing, business, finance, and manufacturing.

Recognition of his abilities was not long in coming. Boeing nominated Carter to participate in the MIT Sloan School program for senior executives, the equivalent of Command & General Staff Officer School for military types. And like Army War College graduates, Carter kept moving up as his experiences broadened.

From program manager for the Air-Launched Anti-Satellite (ASAT) program of the 1980s, Carter moved up to general manager of Space Defense Systems in Boeing Aerospace, adding responsibility for Kinetic-energy Weapons, Directed Energy Weapons, and Space Surveillance and Tracking Systems. Under that flag, Carter also served as Advanced Launch Vehicles Program manager and deputy manager for the Space Systems Division, at a time when the public face of space programs—civilian or military—showed very few Blacks at work.

Visible or not, Arlington Carter was a go-to guy. Organizations under his direct supervision were winning major technology contacts during the late 1980s:

- The Lightweight Exoatmospheric Projectile Program (LEAP) and the building of test high-technology kinetic-energy projectiles for the Strategic Defense Initiative Organization.
- The Neutral Particle Beam-Integrated Space Experiment program, a $40-million procurement to fabricate and test a neutral particle beam device in space.

- The Advanced Launch System, a concept-development effort to determine the system configuration for the Air Force's Heavy Lift Vehicle.
- Shuttle C, a concept-development effort to explore system configurations for NASA's version of a Heavy Lift Launch Vehicle.

Space System organizations under Carter's direction were also indirectly responsible for critical support in Boeing winning the $750-million contract for the NASA Space Station procurement. That made Boeing the main contractor for the Space Station Habitat Modules, where the astronauts now live and work. In addition, Carter's team played an instrumental role in winning a contract for Boeing to build another Initial Upper Stage used to ferry an Air Force payroll into geo-synchronous orbit.

Carter continued his public-service activities, often at his own expense. He served as King County (Wash.) personnel board chairman, a member of the executive board of the Seattle Hearing and Speech Center, secretary treasurer of the Northwest Illinois Institute of Technology Alumni Association, president of the Northwest Area NAACP and chairman of the Western region, an executive board member and loaned executive recruiting manager for the United Way, fund-raising chairman of the Boeing Employees Good Neighbor Fund, president of the Boeing Management Association, and a member of the University of Washington Minority Engineering Advisory Board.

"I've been outgoing in terms of voluntary assignments," Carter told *USBE&IT*. "They're not always plums, but I work on the basis they'll pay off in other ways, and they usually do."

Carter says that volunteering for non-official functions within the company, such as United Way, is also a way of identifying with the company culture and displaying skills that might not be revealed in a person's prime assignment.

"Young engineers have to make an aggressive effort to understand the company, its organization chart, and products. Become a part of the fabric of the company and volunteer for activities, which further the company's goals. As a senior manager evaluating people," he said in 1989, "I'm always looking for the go-getter, the self-motivated person who is interested in the company. Perhaps things like this are not pointed out in brochures, but they do enhance your potential for advancement."

The enhancement and benefits to the community, as attested by the many letters from community organizations supporting Carter's nomination for Black Engineer of the Year, were enormous. And his career success continued apace.

In 1988, Carter was appointed vice president of Boeing Aerospace and assigned as general manager of the Defense Systems Division, responsible for missile systems, Strategic Defense Initiative systems, and support services. Among the major programs he led were the Airborne Optical Adjunct system, the Air-Launched Cruise Missile, the SRAM II strategic missile, and the Avenger and Sea Lance missile system programs.

In April 1989, Boeing merged two divisions to form Boeing Aerospace and Electronics. Carter became vice president and general manager of the Missile Systems Division, adding new responsibilities for the Peacekeeper/Rail Garrison, Minuteman Missile Support and system improvements, and Small ICBM Hard Mobile Launcher-basing system.

At his level, Carter not only had to master and keep up with state-of-the-art space technology, but also the art of political persuasion, a skill he began honing at the Seattle Housing Development Corp. Carter dealt with a wide range of people, within Boeing and in the national polity, from members of the national press corps to members of Congress.

To get there, Carter worked hard to balance his work life with his home life. He told *USBE&IT* that when he first started working at Boeing, he saw that successful executives maintained excellent physical and mental health.

"I felt there were things you committed to the job, but also things you committed to home, community activities, and to leisure," he said. "I saw it as a lifestyle that resulted in prodigious work output, seemingly without effort."

In 1990, when he was named Black Engineer of the Year, Art Carter was vice president and general manager of Boeing's Missile Systems Division. An electrical engineer, Carter spent his entire career advancing with Boeing, working his way up the ranks of leadership. Retired since 1998, Carter serves on the boards of numerous organizations in and around Seattle, including the Citizens Oversight Panel of Sound Transit, a $4-billion transportation improvement project for the Puget Sound area.

Chapter 5

Col. Guion Bluford
1991 Black Engineer of the Year

Guion Bluford was educated at Philadelphia's Overbrook High School—alma mater of basketball great Wilt Chamberlain—and Bluford's 1960 class also included the future NBA All-Star and coach Walt Hazzard, who headed for stardom at UCLA while Bluford went on to study aerospace engineering at Pennsylvania State University. John F. Kennedy was running for president of the United States, promising a "New Frontier" in the fight to extend American democracy, and inspiring young people everywhere to dedicate themselves to public service with the line, "ask not what your country can do for you. Ask what you can do for your country."

Demonstrating its arrival as a power with global reach, the U.S.S.R. sent Sputnik I blasting into orbit in 1957 while Bluford was in high school, besting efforts by U.S. agencies to send up the first artificial satellite and frightening many Americans.

In 1958, the United States answered with a Jupiter-Redstone missile, blasting into space to install Explorer I into low Earth orbit, demonstrating that the Soviets were not the only ones who could send brawny rockets around the world. Kennedy, now president, declared in a speech that Americans would be first to reach the moon, ramping up the priority of aerospace advances.

The space race was on, and many in Bluford's generation harbored fears that competition in building and testing big missiles would end in a nuclear holocaust. Defense budgets soared, and NASA grew into a giant agency with its own billion-dollar budget priorities and installations around the world. Bluford, an Air Force ROTC candidate and future rocket scientist, was in college preparing to become a warrior.

In 1964, Bluford finished at Penn State as a distinguished ROTC graduate and began his active service with flight training at

Williams Air Force Base in Arizona. Pilot's wings in place in January 1965, he transitioned to jet aircraft, qualifying in the McDonnell Douglas F-4C at air bases in Arizona and Florida.

Assigned to the 557th Tactical Fighter Squadron, Bluford flew 144 combat missions out of Cam Ranh Bay, Vietnam. As the war progressed, Bluford flew into increasingly hazardous territory, completing 65 of his missions over North Vietnam. Unlike less fortunate comrades who got shot down and became guests of the infamous "Hanoi Hilton," Bluford got each of his planes back to base in one piece.

U.S. policy is to return experienced combat pilots home to train new fighters, a practice begun in World War II. Not only does this prevent loss of capable air warriors, but it also ensures that they pass on their knowledge. In July 1967, Bluford joined the 3630th Flying Training Wing at Sheppard Air Force Base in Texas as an instructor pilot in the T-38A jet trainer. He flew as a standardization/evaluation officer and as an assistant flight commander for four years.

Like so many others, Bluford watched the news with horror when Martin Luther King was killed by a gunman in Memphis in April 1968, and a short time later, when Sen. Robert F. Kennedy was assassinated in California.

But the next year he also watched Neal Armstrong step out onto the moon on live TV, and the triumph was complete: the Soviets had beaten Americans to the first "soft landing" on the moon with a robot, but Americans had landed the first man. Kennedy was dead, but his promise had been fulfilled, inspiring space dreams in young people the world over.

In early 1971, Bluford attended Squadron Officers School and returned to his Texas base as an executive support officer to the deputy commander of operations and as school secretary to the air wing. Then he went back to school one more time.

In 1972, Bluford entered the Air Force Institute of Technology residency program at Wright-Patterson Air force Base near Dayton, Ohio. Graduating with a master's degree in aerospace engineering in 1974, he won assignment to the Flight Dynamics Laboratory at Wright Patterson as a staff development engineer. He served as deputy for advanced concepts for the Aeromechanics Division and as chief of the Aerodynamics and Airframe Branch of the facility, part of the famed Air Force Research Laboratory. The Vietnam War had ended for Americans in 1973, and Bluford was moving forward as a master of the highest technology: aerospace.

Reports on the Air Force Research Laboratory's work may use fancy, esoteric words, but such research affects much of what we do down on the ground. Among other innovations during the 1960s, Linwood C. Wright, a Black American, developed an efficient high-bypass turbofan engine, a refinement that made turbojet power practical for civilian aviation while dramatically boosting the output power of military jet engines. Turbojets, introduced by the Germans during World War II, had revolutionized powered flight, but were wasteful of fuel. Most of the energy produced went out the tailpipe as waste heat.

Wright figured out how to capture and use that energy: add an extra turbine into the exhaust stream to power a big, ducted fan up front, then use the air stream blasting out from the fan for propulsion. The new efficient turbines quickly put an end to the two-track system whereby jet-setting movie stars and the wealthy swept through the skies in Boeing 707s while less well-heeled passengers took the boat or the train. Like many other American aerospace innovations, it was an advance copied all over the world.

Expanding his own knowledge base, Bluford wrote and presented scientific papers in computational fluid dynamics, critical to rocketry as well as to improvements in powered air flight at the Air Force Research Lab while continuing his studies to complete a Ph.D. in aerospace engineering, with a minor in laser physics.

That moved him up to the front ranks in space exploration. Col. Bluford, who logged 4,000 hours in jet planes, became a NASA astronaut in August 1979, working with engineering systems for the then-new space shuttle, which had not yet flown. These included the remote manipulator system, Spacelab-3 experiments, the Shuttle Avionics Integration Laboratory, and the Flight Systems Laboratory.

Four years later, the name and face of Col. "Guy" Bluford became famous all over the world when he flew on the Space Shuttle

Challenger for three days, deploying the Indian communications satellite INSAT-1B on NASA Mission STS-8.

The Soviets had sent the first Black man into space, Cuban Col. Arnaldo Tamayo-Mendez, aboard Salyut 6 in 1980. The United States had missed its chance to be first with Air Force Col. Robert H. Lawrence, who died in a tragic crash when his F-104 Starfighter plunged from the skies during a training exercise on December 8, 1967. Still, the prestige of the American space program, which was first to land men on the moon and followed that with the first reusable space ship, was undeniable. Guy Bluford, a Black American, was up in the shuttle, and the ability of America's broadcast news media to put color TV pictures in homes, schools and offices in countries on every continent made all the difference in the world. The Soviets were never able to capitalize on their success in beating the United States to yet another space milestone.

Col. Bluford flew into orbit again on Mission STS-61A, a seven-day flight that deployed the German D-1 Spacelab mission and launched the Global Low Orbiting Message Relay Satellite in October 1985. That was the last time the shuttle Challenger went into space before its destruction in a fireball early in 1986. Caused by a malperforming O-ring seal on the strap-on Solid Rocket Booster that let hot gases burn into the shuttle's external fuel tank, the Challenger's loss and the death of Ronald McNair, another Black astronaut, schoolteacher Christa McAullife, and five NASA cohorts badly shook the nation's confidence in the competence of NASA, the world's premier space exploration agency.

Bluford, interviewed by *Pittsburgh Press* reporter Eleanor Chute at a public appearance in May 1989, expressed confidence that the United States would recover its balance in space.

Delivering a keynote speech at the 40th International Science and Engineering Fair held in 1989, Bluford told 746 student contestants from high schools in 46 states, the District of Columbia, American Samoa, Guam, Puerto Rico, and nine foreign nations that being an astronaut was the "best job in the world." But he also sounded an alarm:

"There's going to be a shortfall of scientists and engineers in this country in the years to come," he told a press conference during the competition. "If we're going to be competitive, the leaders in developing new ideas and new products, we're going to need more scientists and engineers."

That was three years after the Challenger's horrific fall. Nonetheless, Bluford's confidence in NASA proved well founded. He went into space again as payload commander for STS-39, an unclassified Defense Department Shuttle flight, in February 1991. Col. Bluford supervised two mission specialist astronauts and monitored all mission-related activities including crew training, payload development, testing and integration of payload gear, mission planning and flight design, flight operations and payload, and flight safety. Col. Bluford also served as point of contact for crew-related issues with the Strategic Defense Initiative Organization, a Reagan-era program to develop space-based anti-missile defenses.

Bluford, now in civilian clothes but still deeply involved with space science, has had a career that spanned nearly the entire arc of the superpower competition. Like the U.S. military-industrial complex that won the Cold War, he continues to contribute.

Chapter 6

Arnold Stancell
1992 Black Engineer of the Year

Arnold Stancell majored in chemical engineering at the City College of New York. Like his contemporary John Slaughter, Stancell grew up without knowing much about Black scientists and engineers who came before.

In 1958, Stancell graduated with his B.S. in chemical engineering and moved up to the Massachusetts Institute of Technology.

Four years later, Stancell began his own career of "firsts." He was MIT's first Black doctor of chemical engineering. During a 10-year stint at Mobil Research, Dr. Stancell developed nine patented processes for making plastics. Still in his research mode, Dr. Stancell opened up a new field of investigation into plasma reactions. He rose to manager of chemical process development, then took a short break to return to his alma mater. On leave from Mobil, Dr. Stancell served as associate professor of chemical engineering at MIT for the academic year 1970–1971. He enjoyed early success mentoring and inspiring a doctoral student, David Lam, who later founded Lam Research, leading maker of plasma etchers for the making of computer chips. Impressed, MIT offered him a tenured post, but Dr. Stancell was headed for bigger things as a manager back at Mobil.

In 1975, NOBCChE, the National Organization for the Professional Advancement of Black Chemists and Chemical Engineers, recognized Stancell's accomplishments with a Professional Achievement Award.

Working as the general manager of Mobil's plastics business, Dr. Stancell revolutionized the packaging industry with a clear plastic that totally replaced the ubiquitous cellophane. He also developed the plastic base for PVC pipes, a vastly cheaper competitor to copper for indoor plumbing.

Dr. Stancell also played a key role in the communications revolution. His research in thin-film deposition greatly affected the computer industry, and one of his plastic products became the "cladding," or outer coating for optical fibers. With cladding that has the proper refractory characteristics, most of the energy in the light beam stays within the fiber. This enables the tiny laser beams used in optical fiber communications to travel thousands of miles, still carrying intelligible signals.

In Europe, Dr. Stancell emerged as a regional manager of marketing and refinery operations, dramatically improving efficiency while learning everything he could about how to put together international deals.

Back in the United States, Dr. Stancell became vice president of Mobil's domestic oil and natural gas business. Then it was back overseas again, as vice president for Exploration and Production of Petroleum and Natural Gas for the Middle East, Europe, and Australia, with a staff of 5,000 employees and a capital budget of $500 million. In this arena, small-timers need not apply. Dr. Stancell, up to the task, initiated, negotiated, and successfully signed a joint venture with the Persian Gulf state of Qatar for natural gas production in the Gulf's North Field, valued at $18 billion during the early 1990s.

Retiring from Mobil Oil after a 31-year career, Dr. Stancell joined the faculty of the Georgia Institute of Technology as professor of chemical engineering, focusing on polymer and petrochemical processes as well as plasma reactions in microelectronics processing.

Over a decade-long career at Georgia Tech, Dr. Stancell won many honors. In 1993, at the end of his Mobil career, the City College of New York recognized him with a Career Achievement Award, and two years earlier he had been the Invited Marshall Lecturer at the University of Wisconsin. In 1997, he was inducted into the National Academy of Engineering and won the American Institute of Chemical Engineering's National Award in Engineering Practice. During the same year, Dr. Stancell was named Outstanding Teacher in Chemical Engineering. In 2001, Georgia Tech named Dr. Stancell chair of Servant Leadership, a position created to "ensure that all undergraduates have exposure to leadership." The chair of Servant Leadership's job, in addition to regular teaching duties, is to teach theoretical courses on leadership and to arrange opportunities for students to apply the principles learned in extracurricular activities.

Stancell was the ideal first tenant of such a chair. Throughout his career, leadership had been his mantra when speaking to young people. Georgia Tech hired him as a teacher of chemical engineering, but it might just as easily have hired the man who affected so many lines of business as a business professor, teaching the principles of executive leadership.

"The world needs technically savvy leaders," said Dr. Lee Wilcox, Tech's vice president of student affairs. And "with the number of organizations available on campus, Tech is a great place to work on leadership."

Leadership on any college campus is usually seen as reserved for select students. Under the initiative by Georgia Tech President G. Wayne Clough, however, the program was designed to provide leadership training and opportunities to a broader number of students. Dr. Stancell's job would be to add a service-oriented tone to the initiative. "The best leaders are service-oriented," Dr. Wilcox told *Technique*, the college newspaper. "They're not there just for the power. They are there to serve."

Dr. Stancell had his own spin: "Servant leaders do not necessarily have to be volunteers. They need to serve the organizations they lead."

Interviewed by the student newspaper, Dr. Stancell talked about the lack of emphasis on leadership in his own professional career.

"None of the engineers in my generation even mentioned leadership. If you were smart and bright, that was it."

Coming from a generation that had been inspired by John F. Kennedy and Martin Luther King, Dr. Arnold Stancell could not escape thoughts of the requirements of leadership. When he realized the need for a leadership emphasis among engineering students, he resolved to "get some leadership" into every course he taught.

Discussing Dr. Stancell's selection as first Servant Leadership Chair, Dr. Wilcox noted that "he had been emphasizing leadership components into his courses, including thermodynamics. He realized through a full career at Mobil that leadership was required to be succsssful."

Thus, Wilcox said, Dr. Stancell "was a natural choice, serving to infuse leadership concepts into the Georgia Tech curriculum as well as developing a stand-alone focal course on leadership, first offered in the fall of 2002. Dr. Stancell also was charged with working to make leadership issues "resonate in different parts of the Georgia Tech curriculum. He worked to infuse leadership materials into courses ranging from freshman-level to major-oriented classes.

"We would like to start emphasizing leadership right from the beginning of a student's stay at Tech and hopefully reach a vast majority," he told *Technique*. "There is a leadership piece in Psych. 1000," Stancell said, referring to one of the required courses for freshmen. He pushed to enhance that portion of the course, and to integrate leadership training into second-year history courses. "American history should include a discussion of historical figures and show good and bad leadership skills in history," he said.

In addition, the Organizational Behavior course, required for industrial engineering and management majors, would expand to cover qualities of leaders in organization. Typical for him, Stancell did not just sit on a committee to create new requirements. Instead, he undertook to personally help the professors in the various disciplines learn how to teach leadership by emphasizing a leadership style that empowered those working under them.

Now retired from Georgia Tech, Dr. Stancell remains active as co-chair of the Board of Chemical Science and Technology, a body established under the National Research Council, which advises the U.S. government on technology issues. Dr. Stancell also sits on advisory boards for chemical engineering at MIT, Carnegie Mellon University, and the City College of New York. Not bad for a kid from Harlem High.

In 2004, *Science Spectrum* magazine named Dr. Stancell one of the "50 Most Important Blacks in Research Science," recognizing his lifelong work in making science part of global society.

Chapter 7

Dr. James W. Mitchell
1993 Black Engineer of the Year

THE FOLLOWING ARTICLE, WRITTEN BY MICHAEL F. KASTRE, WAS FIRST PUBLISHED IN US BLACK ENGINEER & IT, *CONFERENCE ISSUE 1993.*

James Mitchell joined Bell Labs as a member of the technical staff in 1970 after receiving his bachelor's degree in chemistry from the Agricultural and Technical State University of North Carolina at Greensboro and a doctorate in analytical chemistry from Iowa State University at Ames. Later, he accepted the position of supervisor of the Inorganic Analysis Group, and in 1975, he was promoted to head of the department. Mitchell currently heads the Analytical Chemistry research department of Bell Laboratories.

According to Robert A. Laudise, director of the Materials and Processing Research laboratory, "In the past year alone, he has been responsible for a number of important accomplishments that contribute to the increased global competitiveness of the U.S. electronics industry. Using the concept he invented, 'on-demand' reagent generation, which integrates turn-key chemical synthesis with real-time purification and online analysis, dangerously toxic arsine has been produced at precisely determined part per billion levels in order to produce the highest quality silicon wafers for device manufacture. A major supplier of electronic reagents is negotiating terms for the commercialization of this analytical process system."

Mitchell explains the impact on the industry: "In telecommunications, one needs materials that are extraordinarily pure. Silicon was the first example of a broadly used material where it couldn't have trace impurities. The modern equivalent of that is optical wave-guide materials that must be even purer than the specifications we have for silicon. If the optical wave guide materials are not pure, the light pulses that you attempt to transmit through the glass fibers to carry the telecommunications will be absorbed by the impurities."

He adds: "I am really pursuing a line of materials engineering where I am looking at *in situ* generation chemistry and fabrication of thin films with extraordinarily high purity. When one takes excruciating care to prepare materials in states of purity that exceed those that existed previously, you find that they have extraordinary proper ties. Materials that were opaque will now be very transmissive. Materials that were easily decomposed by high voltage are now very stable. Or materials that normally were non-conducive, now become conducive by eliminating impurities."

Mitchell's personal research is one of the cornerstones of modern trace analysis, and his work, together with that of his group, has been absolutely vital to understanding and manipulating the chemistry of electronic and optical communication materials. Without his remarkable analytical chemistry research, neither optical fibers, nor high-purity semiconductors would have advanced to their present stage.

"Substantial quality and bottom-line economic improvements of optical fiber technology is accruing from remedies resulting from Mitchell's process analytical team's diagnostics," says Laudise. "This practical work is exceptionally important in view of the need to greatly enhance the global competitiveness of the U.S. optical fiber industry."

Mitchell has also directed a diamond material program. Comprehensive characterization protocols, including new thermal conductivity methods and infrared luminescence techniques, permit polycrystalline diamond materials to be quality-assured for use as heat sinks for high-power laser devices. He has been granted a key patent for the selective patterning and nucleation of diamond and has initiated new research in carbon vapor transport methods to further improve the properties of diamond.

As a result of his global objectives, he has established "firsts" and received various corporate and professional awards. He is one of only a handful of Blacks to be inducted into the National Academy of Engineering and the first Black to be made a Bell Labs fellow. In addition, he has received the Pharmacia Industrial Analytical Chemistry award, the Percy L. Julian Research award, and two IR-100 awards. Mitchell has also lectured internationally. In addition, he co-authored a book called *Contamination Control in Trace Analysis*, published more that 60 scientific papers, and invented instruments and processes. Mitchell also served as a member of the editorial advisory boards of three international analytical chemistry journals.

A native of Durham, N.C., Mitchell has a love of chemistry that goes back to his high school days. While participating in a college program for high school students, he says, a chemistry professor with a doctorate " showed me a mathematical way of balancing equations." This was something his high school chemistry teacher had limited knowledge about. I was just fascinated by the fact that there was order to chemistry. You didn't have to memorize things. You could figure it out based on established rules. That demonstration showed me that chemistry was also math. So I wanted to learn as much about chemistry as I could." The rest, of course, is history. Mitchell has excelled on the leading edge of technology at one of the world's most prestigious research organizations.

Through persistence and a dedication to education and research, he was able to rise to his current level of excellence. The eldest of five children, Mitchell is the only son of tobacco factory workers. He attended segregated schools and says, "In the era that I grew up in the South, parents did not have to be well educated to prepare their children to be well educated. All the parents had to do was send a disciplined child to school and support them when they came home." The standards and efforts of this remarkable engineer should inspire all youth to aim high and strive to be the best. "Once I earned a doctorate degree, my decision was that I wanted to be known in the circle of the best people on earth who work in my field," Mitchell says. He has exceeded that goal. Mitchell and his wife Jean have three children. One is in medical school, another studies microbiology, and the youngest is in high school.

Chapter 8

Dr. William Wiley
1994 Black Engineer of the Year

THE FOLLOWING ARTICLE, WRITTEN BY MELINDA COOKE, WAS FIRST PUBLISHED IN US BLACK ENGINEER & IT, *CONFERENCE ISSUE 1994.*

Bill Wiley has spent a lifetime figuring out how to make technology work for people. Dr. Wiley is the principal executive in the Northwest for the Battelle Memorial Institute, an independent, worldwide, science-based organization dedicated to "putting technology to work."

Wiley's responsibilities include research, development, and technology commercialization at the U.S. Dept. of Energy's Pacific Northwest Laboratory, as well as Battelle's Marine Sciences Laboratory in Sequim and the Seattle Research Center.

Wiley, 60, who holds a doctorate in bacteriology, supervises more than 4,000 scientists, engineers, and technical specialists doing research in support of DOE missions in environment, energy, defense and national security, economic competitiveness, and education. The lab is also participating in the 30-year, $1-billion per year cleanup of the nearby Hanford site.

Born in Oxford, Miss., Wiley graduated from Tougaloo College with a degree in chemistry. After serving in the U.S. Army, he attended the University of Illinois at Urbana under a Rockefeller Foundation scholarship. At Illinois, he received his master's degree in microbiology in 1960. He then went on to obtain a doctorate in bacteriology from Washington State University in Pullman. He started his career with Battelle in 1965 as a research scientist at the Pacific Nothwest Laboratory. In 1979, he was promoted to director of research, and occupied that position until 1984, when he became laboratory director.

Wiley is presently responsible for the business operations of Battelle's Pacific Northwest Division, which has operated PNL for DOE since 1965. As a result of his vision, PNL is becoming a major center for science and technology in the Northwest. Wiley feels that the future of this country's economic growth needs to come from research institutions like PNL. Many leading experts in infrastructure and economic policy agree that scientific research facilities like the DOE national labs will play a major role in America's economic development. "The national labs have an existing research infrastructure that is ready to contribute advanced technologies for the seven or eight new industries that will be the basis of future economic growth," said Wiley in an interview following his participation in President Clinton's fall 1992 economic summit.

As a boy in Mississippi, Wiley had dreams of becoming a doctor. He realized at an early age that this vision could only be achieved through education. He continues to believe that investment in education is one prescription for America's healing. "Think of it in business terms," Wiley says. "It's a long-term investment that is bound to result in a net gain.

To this end, Wiley has served on many local and regional task forces aimed at developing long-term educational goals for the minority community. To encourage more minority students to pursue careers in science, he helped create programs at PNL that will support more than 100 African-American students this year. These students will participate in education and research activities at the lab under the mentorship of experienced scientists and engineers. Wiley also puts his philosophy to work by sending Battelle's scientists and engineers into classrooms to be role models to students. Through his Adopt-a-Teacher program, teachers can spend a summer observing and participating in research projects and can take their newfound knowledge back to their students.

His solid reputation as a scientist and an administrator has led many groups to seek Wiley's advice. He was selected to join the 61-person group helping Washington governor Mike Lowry to plan for his administration. Wiley is the first national laboratory director to be a member of the Council of the Government-University-Industry Research Roundtable, which reports directly to the presidents of the National Academies of Sciences and Engineering and the Institute of Medicine. He is president of the Washington State Board of Regents and a member of the Fred Hutchinson Cancer Research Center Board of Trustees, and the Southern University Engineering Executive Committee.

As an award presenter at last year's Black Engineer of the Year Awards conference, Wiley saw firsthand its benefits to the Black community and the country at large. "The real value of the Black Engineer of the Year Awards," he says, "is the impact that it must have on students and young professionals starting their careers in very large corporate organizations. The conference is important to the country because it promotes visible demonstrations of how diversity in the work place contributes to the bottom line of American companies." But it is in the Black community that he sees its most important effect. "Wide distribution of the proceedings to counselors of high schools," Wiley says, "in predominantly Black communities, could give help and guidance to the careers of otherwise delinquent youth." Wiley's vision to tap into America's education and research resource facilities and stimulate economic growth by forming an alliance with cutting-edge industries is now embedded in the mainstream business philosophy that has been endorsed by President Clinton. It seems as though the Mississippi boy who once wanted to write prescriptions may have helped pioneer a cure for some of America's most pressing economic ills. Innovation is the true mark of this pioneer's life and works, one that makes him the embodiment of the Black Engineer of the Year.

DR. WILLIAM WILEY, MICROBIOLOGIST, EDUCATOR, BATTELLE MEMORIAL INSTITUTE VICE PRESIDENT, AND BLACK ENGINEER OF THE YEAR IN 1994, WAS "A LIFELONG BELIEVER IN THE ABILITY OF RESEARCH ACTIVITIES TO DRIVE ECONOMIC DEVELOPMENT AND, THROUGH IT, TO CHANGE PEOPLE'S LIVES." AS THE PRINCIPAL EXECUTIVE IN THE NORTHWEST FOR BATTELLE—SUPERVISING 4,000 SCIENTISTS, ENGINEERS, AND TECHNICAL SPECIALISTS—HE DEVOTED HIS ENERGIES TO FIGURING OUT HOW TO MAKE TECHNOLOGY WORK FOR PEOPLE. DR. WILEY DIED AT HIS HOME IN RICHLAND, WASH., ON JUNE 30, 1996.

Chapter 9

Walt Braithwaite
1995 Black Engineer of the Year

The following article, written by Travis E. Mitchell, was first published in US Black Engineer & IT, *Conference Issue 1995.*

Boeing's Walt Braithwaite is sitting on top of the world these days. In his current capacity as vice president, Information Systems, Braithwaite is responsible for all information systems activities for Boeing's Commercial Airplane Group. As the functional head, he is responsible for some 4,000 people and an annual budget in excess of $1 billion. According to his job description, "he has to make sure that the computing resources and infrastructure necessary to support design, production, and in-service support of all airplanes is in place." In layman's terms, it means that the commercial airplanes that Boeing produces won't make the cut unless he gets the design, computer technology, and resources in place on the front end. Also, he has to ensure that these component parts stay in place during the entire production process—no small task.

Braithwaite helped perfect Boeing's use of computer technology in the design and manufacturing process. Through the use of computer-aided design, engineers now are able to assemble entire planes down to the smallest bolts, and trap out any flaws before the actual model is built. These innovations have saved millions of dollars in man-hours by cutting the research and development phase in half.

"As an engineer, Walt has made innovative contributions to Boeing and the engineering community," says Frank Shrontz, the Boeing Company chairman and chief executive officer. "His work in computer-integrated technology was crucial to the development of the CAD/CAM systems that now make it possible for us to design airplanes by computer."

"Since Walt joined Boeing, he has contributed to programs essential to the changes we are making in the way we design and produce airplanes," says the Boeing Company President Phillip M. Condit. "The importance of his technical contribution to computer-integrated technology extends beyond Boeing."

In 1985, Braithwaite received an award in recognition of outstanding effort and guidance in the creation of the Initial Graphics Exchange Specification. In 1987, he received the Joseph Marie Jacquard Memorial Award presented by the American Institute of Manufacturing Technology for outstanding technical contributions to the science of computer-integrated technology. More recently, in 1990, he received the Product Definition Exchange Specification Award presented by the IGES/PDES Organization for leadership and outstanding contributions to the development and production of the Initial Graphics Exchange Specification.

"We came up with the standard to allow different computer-aided design systems to communicate," says Braithwaite. "We did it at Boeing first and I took it to the Academy of Sciences in Washington, D.C., where we discussed it with other people from industry. As a result, we decided to use it as the foundation from which to develop a standard."

For his outstanding technical achievements and trailblazing career path at Boeing, Braithwaite is this year's Black Engineer of the Year.

"This guy is amazing," says Garland Thompson, editor of *Black Issues in Higher Education*, and a member of the Conference's awards selection committee. "What stood out in his package was the fact that he helped to set industry standards for CAD, which today is a commonplace tool in technical industries. He is a trailblazer in every sense of the word."

Briathwaite has made patience his home formula for success.

"My mother did a lot of different things," says Braithwaite. "She was a beautician, seamstress, and embroiderer. My father was laborer. He would repair shoes and do cabinetwork with only formal apprenticeship training. They always worked hard. It was just natural for me to adopt that type of work ethic."

Braithwaite knew early on that he was not going to subscribe to the contemporary thought that he couldn't achieve because of his skin color. Rather, he employed a can-do or must-do attitude toward education.

"Those of us who come from places like Jamaica, where you had to work hard to make sure that things happened, saw all these opportunities; we just had to reach out and take them," says Braithwaite. "It's like a chess game: when you are in the game, you can't see all of the moves. Somebody standing on the outside looking in can see the moves. So coming to the United States, you see the opportunity and you say, 'Wow! I didn't have that at home so I'm going to take advantage of that here.'"

Braithwaite followed his childhood dreams and received a Bachelor of Science degree in engineering from the American Institute of Engineering and Technology in Chicago, Ill., in 1965. Shortly thereafter, he joined Boeing as an associate tool engineer in the Fabrication Division in 1966. In 1975, Braithwaite moved to the engineering payroll as a senior engineer responsible for providing technical direction in the development of the CAD/CAM Integration Information Network, and the Geometric Data Base Management System, which became the foundation for Boeing's CAD/CAM activities.

Braithwaite's work in the development of a standard data format for the exchange of digital product definition data between different types of systems established the basis for the American National Standard. His work in that area continued as he became the lead engineer in Boeing's Everett Division, where he directed computer-aided design and development and was the chief engineer responsible for computer-assisted design and manufacturing. During this time, Braithwaite continued his postgraduate studies in computer science at the University of Washington in Seattle. He completed his degree requirements in 1975 and then began to show interest in management. And management showed interest in him.

In 1980, Boeing sent him to the Massachusetts Institute of Technology as a Sloan Fellow in business management. Condit says that the company offered the fellowship "in recognition of his achievements and his future promise as manager."

The appointment, no doubt, was helped along by his technical innovations.

"In the late 1970s, the interactive design systems were becoming a reality. Because of my training, continued studies, and awareness of the new computing technology and developments in industry, I was placed in a position where I could contribute significantly to the Boeing Company," Braithwaite says.

In 1985, Braithwaite was named director of computing systems in the Everett Division, where the 747 and 767 aircraft are assembled. From 1986 to 1991, he was director of program management for the 737s and 757s in Renton, Wash. In 1991, Braithwaite was named vice president, Information Systems and Architecture, and in May 1994, he assumed his current post as vice president of all information systems activities for the Boeing Commercial Airplane group.

"When I joined the company in 1966, I saw people getting their five-year pins," Braithwaite recalls. "I didn't think that I would be here that long. I never thought I would get one of those pins. But once that happened, I started to set goals, and I've been here now for more than 28 years."

"I have had a professional association with Walt for more than 12 years," says Ronald B. Woodward, president, Boeing Commercial Airplane Group. "As a member of the Boeing Commercial Airplane Group Leadership team, Walt plays a key role in focusing our direction to remain number one in the world. I value his decisions on the use of information systems, as they are directly related to the achievement of our production strategy. He is clearly our resident expert in this arena."

Braithwaite knows that his family is the backbone of his career success. He credits his wife, Rita, with that. She has been intimately involved with the rigors of handling the children's academic preparation, while her husband puts in the long hours in the office.

"Rita has been there on the front lines the entire time," says Braithwaite. "While the children were in elementary and middle school, she was actively involved in the P.T.A. and other parent groups. She has been the glue that has held us together."

Rita and Walt have two children: Chrissy, a ninth grader, and Cathy, a freshman at New York University. Walt has another daughter, Charlene, who lives and works in New York.

"I try to make it so that his life is not all work," says Rita. "When Walt comes home, I try to get his mind off work so that he can relax.

Not having work is the hardest task for this man, who is so dedicated to his job that some who know him have tagged him "Sleepless in Seattle."

When Braithwaite does find time to wind down, he usually spends it sailing his boat or flying. Not only does the man know how to design aircraft, he also knows how to fly them. He is a true Renaissance man. Braithwaite totally refurbished his boat from what Rita called "junk" to a prize-winning vessel.

Dedication is one Braithwaite trademark that he leaves on all of his projects. He says that it took him two hours a night for two years to finish restoring his boat. He has dedicated the same energy to a local YMCA Black Achievers program.

About Braithwaite's dedication to youth, former Black Engineer of the Year, A.W. Carter Jr., the Boeing Company's corporate vice president of facilities and continuous quality improvement, says, "He is an extraordinary role model for young African Americans, both those who should be encouraged to pursue an engineering career, as well as those aspiring youngsters presently in engineering curriculums in our Historically Black Colleges and Universities."

"Walt has demonstrated a willingness to concentrate his energies on activities and projects that motivate young people to reach for something better," says Seattle's mayor, Norman B. Rice. "I am sure that as Walt's career has blossomed at the Boeing Company, constraints on his volunteer efforts have increased. However, he has made young people a priority in his life."

Braithwaite even reaches out to his native Jamaica, providing schoolbooks and other educational materials to the elementary school that he attended as a boy. "Once while I was visiting my old elementary school, I saw a kid hammering a nail into a desk with a rock and I asked the teacher what he was doing," says Briathwaite. "She told me he was trying to fix his desk. Since that day I have tried to do what little I can, whenever I can. ... I go to my kids' schools and see all of the nice computers and labs, and it really puts life into the proper perspective. I try not to let my kids forget the privilege that they have. Similarly, I am quick to remind the youngsters that I tutor that education is a privilege and an opportunity that they must use to their advantage. I try not to get preachy, but I feel it's my duty to share the keys to my success with young people, so that they might soar to new heights."

Chapter 10

Lt. Gen. Albert J. Edmonds, USAF (Ret.)
1996 Black Engineer of the Year

THE FOLLOWING ARTICLE, WRITTEN BY GRADY WELLS, WAS FIRST PUBLISHED IN US BLACK ENGINEER & IT, *CONFERENCE ISSUE 1996.*

It is almost a truism that successful modern warfare is information intensive. Soldiers, who can see farther in chaotic and difficult situations, can project their strengths before their enemies can. But the same up-to-the-second information, which provides advantages to military commanders, is vital to civilian leaders in times of crisis. That is why the Defense Information Systems Agency is in charge of federal emergency communications. The same built-in reliability, clarity, and wide reach necessary for troop communications under battlefield conditions are critically needed by the leaders of federal agencies when disaster strikes. The parallels are striking.

During the Gulf War, for example, global positioning satellites allowed coalition forces to navigate through the forbidding deserts in southern Iraq. Computer networks and radio-linked armored vehicles helped coalition commanders instantly locate Iraqi tanks and troop concentrations. When defeated Iraqi military commanders were shown the small devices that pinpointed our forces' locations, they were astonished that even relatively low-level soldiers had such technology at hand. At home, after a devastating storm such as Hurricane Hugo, when power and phone lines are down, roads are washed out and entire communications are laid waste, similar abilities are needed. Authorities have to locate the scenes of worst damage, rapidly move rescue and cleanup crews into the hardest-hit areas, and open lines of communication to local-government leaders, private relief agencies, and citizens needing help and encouragement.

As director of the Defense Information Systems Agency (DISA) and manager of the National Communications Systems, Lt. Gen. Albert Edmonds is responsible for making sure the best information and communications systems are in place for our forces, at home as well as abroad. His agency provides command, control, communications, and computer and intelligence (C4I) support to civilian authorities as well as to the military's war fighters. The reason is that many systems intersect at DISA, including the Global Command and Control Systems (GCCS), the National Communications Systems (NCS), the Defense Information Infrastructure (DII) and the National Information Infrastructure (NII), the information super highway. In a very real sense, Lt. Gen. Edmonds is building the information superhighway of the future, today.

"The Global Command and Control Systems is the heart of the C4I for the Warrior Initiative," Gen. Edmonds says. "GCCS is the realization of a Joint Staff goal to provide a single, seamless, interoperable command and control structure that uses commercial and government hardware and software that has allowed us to put real-world operational functionality in the war fighters' hands today.

"GCCS was successfully used during Operation Uphold Democracy," he says, referring to the 1994 operation that restored democracy to Haiti. "The secretary of defense and the chairman of the Joint Chiefs of Staff saw firsthand an operational system providing a true picture of the evolving battle space in real time from the Pentagon. GCCS provides our nation's war fighters with a fused, real-time, true picture of the battle space, which will give the war fighters an ability to order, respond, and coordinate, horizontally and vertically to the degree necessary to prosecute our mission in that battle space. GCCS capabilities include crisis planning, force deployment and employment, personnel, logistics, force status, intelligence, positions, and direct operations.

"GCCS takes advantage of military satellite communications, commercial satellite communications, asynchronous transfer mode/synchronous optical network leading edge technology, fiber, and other terrestrial communications capabilities available through the Defense Information Infrastructure."

Putting aside the muscular language of the warrior, the DII is a seamless web of communications networks, computers, software, databases, applications, data, and other capabilities that meets the information processing and transport needs of Department of Defense (DOD) users in peace and in all crises?conflict, humanitarian support, and wartime roles. The DII includes the following:

- Physical facilities used to collect, distribute, store, process, and display voice, data, and imagery.

- Applications and data engineering practices, tools, methods, and processes, to build and maintain the software that allow C2, Intelligence, and Mission Support users to access and manipulate, organize, and digest proliferating quantities of information.

- Standards and protocols that facilitate interconnection and interoperation among networks and systems and that provide security for the information carried.

- People and assets, which provide the integrating design, management, and operation of the DII, develop the applications and services, construct the facilities, and train others in DII capabilities and use. The DII will take many unintegrated, single service "stovepipe" information systems and replace them with a globally distributed, user-driven infrastructure, a so-called Global Infosphere, through which military leaders can gain access, from any location, for all required information. What soldiers will see from the DII is a fused, real-time, true representation of the three-dimensional environment in which a battle must take place.

Gen. Edmonds' other title, manager of the National Communications Systems, gives him responsibility for the planning, coordination, and integration of government telecommunications capabilities that ensure an effective response in all hazardous situations. The NCS' beginnings in the aftermath of the Cuban Missile Crisis tasked the agency with ensuring that the U.S. government has the telecommunications necessary to meet its national security and emergency preparedness (NS/EP) responsibilities under all conditions. The NCS has served in many capacities. It provided critical support during Operations Desert Shield and Desert Storm and helped speed the recovery after the Northridge, Calif., earthquake. It also plays a major role in today's humanitarian-aid efforts throughout the world. The NCS, located in Arlington, Va., brings together 23 government organizations to address the full range of NS/EP telecommunications issues. The NCS is continually evolving to meet new and complex challenges, including the reliance of the federal government and the nation on the public switched telephone network and on the developing National Information Infrastructure.

Environmental forces add to the complexity of these challenges, most notably: declining federal expenditures, evolving threats, electronic intrusion, terrorists, natural disasters, and the rapid emergence of new information technologies. The NCS' mission has, over the past several years, evolved from one of emergency telecommunications response for national security emergencies to an all hazard response capability that covers natural and man-made disasters as well. This evolution came about through dramatic changes in the political, economic, and technical environments, changes in the United States' national security and economic posture, and the reduction in the threat of massive nuclear war due to the decline of communism and the fall of the Soviet Union.

In the wake of the Cold War, the threat to U.S. citizens is more likely to come from a natural disaster than a nuclear war. The NII initiative, as envisioned by the Clinton administration, is intended to exploit advances in information technology to improve national competitiveness, alter the way the federal government delivers services to its citizens, enhance the delivery of health-care services, and expand educational opportunities. In this context, the NCS perspective is to leverage these information technologies to improve the federal government's ability to respond to all hazards. The NCS' role in the evolving NII is to ensure that NS/EP telecommunications requirements are considered in developing standards, defining needed services, instituting policies and legislation, developing acquisition guidelines, and providing research and development fiscal incentives.

Specifically, the NCS is seeking to define and promote NS/EP features that should be available to government users on the NII. These features include priority service provisioning and restoration procedures, emergency broadcast capability, a sustainable NS/EP coordinating mechanism, and assured and reliable service and network security.

Since he has spent an entire career in military communications electronics, Gen. Edmonds has worked with hundreds of engineers, in uniform and out.

"In my current position as head of the Defense Information Systems Agency," he says, "I rely on the expertise of our engineers to build a little, test a little, and field operational systems, implementing the C4I for the warrior vision of a fused, real-time representation of the warrior's battle space. They have assisted in refining the DOD information technology roadmap to clearly show how to achieve progress using a coordinated approach to right size C4I systems using new and affordable technology that is smaller, more capable, and readily able to be deployed with the warrior. We're designing, building, and fielding systems that are interoperable, and we're using commercial and government technology to do it. Everything we do or build has an engineer's fingerprint on it."

For students, Gen. Edmonds has specific advice on preparing for the career of the future, especially in the service.

"Anyone thinking of entering the military," he says, "should focus on the future and pursue majors and electives which will prepare them for the fast-paced, technologically challenging 21st century. The military will need individuals with technical, scientific, and analytical skills, as well as individuals who can identify problems, provide logical, practical solutions, and implement them in a rapidly changing environment. We need individuals who can function comfortably within the military structure and implement change to accommodate new technology. Majors and electives which provide individuals with these skills will be in demand."

Edmonds holds a B.S. in chemistry and a master's degree in counseling psychology from Hampton University. He completed the Air War College as a distinguished graduate in 1980, and Harvard's national security program for senior officials in 1987. He also has an honorary doctor of science degree from Morris Brown College.

"I am indeed honored to be recognized by this prestigious group of engineers," Edmonds says. "Having spent much of my career working with engineers designing and developing networks to support our war fighters, I have come to respect their talents and creative skills in finding innovative solutions to meet the military's technology needs."

Chapter 11

Art Johnson
1997 Black Engineer of the Year

The following are excerpts from the 1997 Black Engineer of the Year Awards ceremony speeches honoring Art Johnson. The theme of the conference that year was "Through the Eyes of a Child."

The Old Storyteller: So, I have but one last story to share with you. Afterwards, it will be in your hands. To live life is to learn and lead by example. Life's true champion is the one who, when growing up, chooses the right leaders to follow and when out front, chooses to lead the followers right. Let us look at one who has hurdled some staggering odds by virtually ignoring their existence and fixing his gaze on his goals. Let me show you what it takes to become the Black Engineer of the Year.

Norman Augustine, vice chairman and CEO of Lockheed Martin Corporation: Since the end of the Cold War, mega-mergers within the defense industry have made executive jobs hard to hold onto. Art Johnson, a group vice president with Lockheed Martin Federal Systems, is one notable exception. Art, who was head of the former IBM Federal Systems Division and a group vice president, stayed at the top when the division was sold to Loral. Another merger swept the group into Lockheed Martin, with Johnson still at the helm. And a recent promotion made him a Lockheed Martin corporate vice president, as well as division president. All along, he has proved to be one of the key players in defense communications.

Art leads a division with nearly 8,000 workers. The company sold over $2-billion worth of goods and services last fiscal year, more than a quarter of which came from foreign sales. While the division was part of IBM during Desert Storm, it provided the decisive communications that worked remarkably well amid the difficult and hostile terrains of the Middle East. The technical elements responded quickly with practical solutions that contributed to the many successes of the allied forces.

Art's entire career has been engineered toward such successes. After taking his degree from Morehouse College, he became deeply involved in the space program for IBM at Johnson Space Flight Center in Houston. He became the administrative assistant to his division's president and Chief Operating Officer, where he learned how everything worked. A decade later, after an executive assistant job with Chairman John Akers, Johnson moved up to lead Federal Systems.

In 1993, President Clinton named Art to a three-year term on the Defense Science Board. He joined the Presidential Advisory Council on Historically Black Colleges and Universities in 1994. As an HBCU graduate himself, he believes these institutions are national treasures and must be nurtured and maintained. Art serves on the board of the Armed Forces Communications Association, and he is a mentor to many small businesses.

It doesn't do justice to his achievements simply to list them. There is a story behind every success and right now let's ask the Old Storyteller for one version of the Art Johnson story.

The Old Storyteller: Arthur Johnson. Oh, yeah, I know him. He grew up in Durham, North Carolina, right down the street from me. I knew his folks, too. When he was growing up here, it could have been hard times. Jim Crow and segregation were designed to lessen one's dignity. For lots of folks I reckon it was hard times. But for that family, it was different—different because they didn't believe in hard times. Least that's the way I had it figured. See, the parents of that boy were two downright pertinacious people. They were never going to take "no" for an answer so long as "yes" had just as many syllables.

His daddy was a school teacher, I think, and his mama ran the head start program over here. He used to tell me that one of the most inspirational moments in his whole life was when he saw his mama walk down that aisle to get her college degree when he was in the eighth grade. Both his parents were college people and it was expected of him. I used to walk past the house from time to time and listen to the family talking. Well, I'd listen to him talking while everybody else listened. Most of what I heard him say was "why?" or "how come?" One question after another! On and on. And his family encouraged him. They wanted him to know. They wanted him to want to know. And they wanted him to know when he didn't know.

Mama used to say, "we need to cultivate his curiosity." I wanted to cultivate something all right, but it wasn't his curiosity. But she was right of course, to let him go on like that. See where it brung him? The whole community where he grew up was like that. They worked hard and talked all the time about learning and thinking. In school, Mrs. Thomas used to tell him he had to learn to analyze the situation and to think logically about the outcomes. Mr. Gaddis taught him to express himself with confidence and resolve. Work hard, learn voraciously, stay calm, cultivate curiosity, and always, always talk to the persons you encounter. This is what he grew up with in Durham. And then when he got to Atlanta, down there at Morehouse College, he ran into Benjamin Mays, one of the premier scholars of this century. Dr. Mays kept up the drumbeat for hard work and commitment. He used to tell all his students to "strive to be successful, but what's more, you must make a difference." This young man took it to heart. That is why he is so involved in the community. That is why he rose to the top in management. He was taught the value of smart work, hard work, and keeping his commitments. Management is a 24-hour-a-day job. And he'd probably say you can't take a 24-hour-a-day pay out of nine-to-five work.

Well, I could tell you more but I won't. You got things to do, and so have I. But if you're wondering how a lad from Durham rose to the top of the heap to manage, well, he had an "I can" mama and an "I can" daddy, and that's a double "I can" advantage.

1997 Black Engineer of the Year Art Johnson: First, let me start by thanking Norm Augustine for being here this evening and presenting me with this award. Norm committed to me some time back that if I was fortunate enough to be selected for this honor, he would like to be here as presenter. He was in Georgia earlier today but he came back to meet his commitment and that is very important to me.

Norm, I am pleased that you could be here and I want you to know that I am proud to be a part of the Lockheed Martin team.

Second, I would like to both congratulate and thank Tyrone Taborn and Career Communications Group. Congratulate them on this tremendous recognition vehicle and the wonderful role they perform and the very professional job they do.

I say this because I believe that more important than the individual recognition that you see here tonight are the reverberating messages sent by this process.

And of course, it is a tremendous honor for me to receive the Black Engineer of the Year Award—and I am also honored and privileged to be able to play a role in this communications process. I would like to thank and acknowledge my friends and colleagues who are here this evening to celebrate this award with me. And then, I would like to give a special thanks to my family—my daughters, Mia and Leslie Johnson, and my wife, Carol Johnson. They are here with me, sharing in this distinctive recognition as they have shared in, sacrificed for, and contributed to all of my accomplishments.

You know, as I think about this conference, I am intrigued somewhat by the juxtapositioning of tonight's awards, focused on engineering and technology, and the theme—"Through the Eyes of a Child." Because unlike my formative years or those of most of you here this evening, the foundation, perspective, and experience base of today's generation of children will increasingly be driven by information technology. It will have a pervasive effect on all aspects of their lives, and it will be an integral part of the language they speak. Even now, we program our VCRs; in our cars, we check the trip computer; we run diagnostics on our kitchen appliances; and need I remind you of the ubiquitousness of the Worldwide Web?

In the past, information technology was critical to our success. In the future, it will be critical to our existence. And that is why this process we are a part of tonight is so important, and why the message communicated by these awards is so vital: vital in reinforcing vignettes of success in the demanding engineering fields; vital in portraying positive images of accomplishments and contributions; vital in giving a view of the tip of the iceberg of challenge and opportunity; and vital in offering a quick peek at the realm of possibility that is at our fingertips.

That is why I am excited to be a part of this process.

Chapter 12

Lt. General Joe N. Ballard (Ret.) 1998 Black Engineer of the Year

The following are excerpts from the 1998 Black Engineer of the Year Awards ceremony speeches honoring General Joe Ballard. The theme of the conference that year was "Reflections: On the Pathways to Success."

General Dennis Reimer, United States Army (Ret.): He came from a small southern town, but his work affects every part of the country and our Army throughout the world.

Peter Teets, Undersecretary of the Air Force: His decisions can affect every American's life.

Gen. Reimer: The U.S. Army Corps of Engineers, which he commands, is the world's largest public engineering, design, and construction management agency. It has a budget of more than $11 billion.

Mr. Teets: The Corps of Engineers maintains 300 deep-draft harbors, 275 locks, and 12,000 miles of navigable waterway. It is charged with keeping the inland waterways dredged and navigable for the passage of the nation's waterborne commerce. And anyone wishing to mount construction in the nation's wetlands must first get permission from the corps.

Gen. Reimer: He manages 383 flood-control lakes and reservoirs and 8,500 miles of levees to help protect people from the ravaging damage of floods.

Mr. Teets: And when the floods do come, like the disastrous Red River floods of the Midwest or the mighty Mississippi floods of last year, his corps mobilizes to manage the river's flow and reduce and repair the damage.

Gen. Reimer: Most people aren't aware that a full quarter of the nation's hydroelectric power comes from 75 facilities run by the corps.

Mr. Teets: Or that this fiscal year, the corps will invest almost a billion dollars to protect the environment.

Gen. Reimer: Nearly all of the lakes and reservoirs managed by the Corps of Engineers are open to public recreational use.

Mr. Teets: And the corps collaborates with other agencies providing reservation services for 20,000 family campsites.

Gen. Reimer: And the military role for the Corps of Engineers is actually larger than the civil works that we have just discussed.

Mr. Teets: The Corps of Engineers is the real estate agent and base construction manager for the Army and Air Force. They build dining facilities, training ranges, barracks, and hospitals for America's soldiers and airmen.

Gen. Reimer: But when the nation goes to war, it is often the engineers who lead the way. They clear mine fields, construct airfields and bridges to bring in the rest of the army, and then they build the camps and facilities to sustain them.

Mr. Teets: So who is this man, this soldier's man who creates the vision for a command so strange and diverse? And who, may I ask, is this engineer, and where's this place from which he came, and wears the golden castles, and presumes he can move the earth?

The Old Storyteller: This is the Ballad of General Joe Ballard, Keeper of the Golden Castles. He was born in Meeker, La., well south of the Mason-Dixon line, at a time when the land was divided against itself. The oldest of six children, he was raised by his mother who managed the at-home work and his daddy, who managed the away-from-home work. When young Joe was four or five years old, his family moved to Oakdale. Oakdale was one of those character-building towns, where hard times were overcome by hard work.

Sounds like a strong place with deep roots, doesn't it? Oakdale was an all-Black, family town. Aunts, uncles, and grandparents all lived just across, or just down, or just up the street, and they, each one, was a surrogate parent.

It was a virtual fortress of values, Oakdale. Everybody took charge of teaching something. His mama and daddy were in charge of character. His daddy was a hard-working man and a Baptist preacher, and his Mama was a full-time mother! They taught by their example the values of fortitude, attitude, gratitude, and bliss. These values stuck to Joe like chocolate to a chubby little cheek.

His Papa—that's his grandfather on his mother's side—lived just across the street and took charge of expectations. He presented a grand image at six feet, four inches tall, manicured mustache, erect bearing, and sharper than a Remington razor. Papa used to call Joe the President of the United States. That's setting the expectations real high for a little Black boy in Louisiana in the 1940s and 1950s, but that was the nature of that town.

Whenever Papa called him president, Joe's eyes would pop out with pride. Then Papa would say, "you gotta be president cause you're too lazy to work!" and he'd laugh that big laugh that only papas seem to know how to do. It was Papa who planted the notion of higher schooling in his imagination. Papa just assumed he was going to go to college. And he was right, Papa was. When Papa died, Joe was a sophomore at Southern University.

Uncle Charlie lived just down the street and took charge of exposure. He was a real fine dresser, too, and had been around the world in the Navy and had stories and pictures to prove it. It was through Uncle Charlie's stories that Joe had adventures around the world before he ever left Oakdale, exposure to wonders that would eventually lead him to the top.

Then there was Mrs. Glenn, his fourth grade teacher. She took charge of education. He said that she is both his most and least favorite teacher. What a compliment. That means she got the balance just about right. It was Mrs. Glenn who gave him a passion for the written word. She was his most favorite teacher because she showed him the magic and wonder that was hidden between the bindings of the books. She was his least favorite when he forgot that first lesson. Mrs. Glenn said to him, "If you read well, there is nothing else that you can't do." Then she introduced him to the library and William Henley's poem, "Invictus." Wow!

Well, I've gone on too long, and we haven't even left Oakdale. I was telling you how he came to be keeper of the castles. So let me finish my story with that.

Legend has it that in 1903, Gen. Douglas MacArthur was given these solid gold castles by his father, symbolizing the building trade of the engineer. They were passed down until the family decided that the chief of engineers in the Army should wear the castles to signify the top office. So at the installation ceremony, the chief is adorned with the golden castles bequeathed by Douglas MacArthur. General Joe Ballard is the first Black man to wear those castles.

That is a good story. But I don't know if you realize that the word castle means a fortress. A built-up entity designed to defend against attack. I told you that Oakdale was a fortress of values. Each value is a golden brick used to construct the most intricate fortress of character that can transport one from Oakdale, La. in the 1940s to the chief of engineers in 1998. So when we say he is the keeper of the castles, he was that long before he became chief of engineers. In fact, he became the chief of Engineers because he was such an exquisite keeper of the castles of character. Perhaps Mrs. Glenn saw it a long time ago when she made him read: "It matters not how straight the gate, how charged with punishments the scroll, I am the master of my fate, I am the captain of my soul." Invictus means "unconquerable," and that one word, my friend, is the Ballad of General Joe Ballard, the Keeper of the Golden Castles.

1998 Black Engineer of the Year Lt. General Joe N. Ballard (Ret.): Thank you General Reimer and Mr. Teets. To accept this award from men of your stature makes it even more special for me.

When I was growing up, the only kind of engineer I had ever heard about was the engineer who drove the train. Then I met a man who changed my life. That man, Mr. Jesse Anderson, owned an electronic repair shop in my hometown of Oakdale, Louisiana. He was one of the few Black businessmen in town. Mr. Anderson could fix anything—radios, toasters, even the new invention of the day: TV. You name it; he could fix it.

When I was in the eighth grade, I started hanging out at Mr. Anderson's shop. He put me to work doing odd jobs and learning to fix things. I discovered that I was not only good at fixing things, but also actually enjoyed it.

Mr. Anderson would talk to me every day about going on to school to become an "electronic engineer," as he called it. He opened up my mind to the idea of studying engineering and paved the way to my future. It is directly because of Mr. Anderson's support and encouragement that I stand on this stage tonight.

Allow me to say a word about the theme of tonight's celebration: reflections. The premise on which this theme is based is that telling the stories of yesterday's experiences galvanizes our strength for tomorrow's encounters. I don't know about you, but after hearing these marvelous stories tonight, and reflecting upon Mr. Anderson's role in my life, I am made more ready for tomorrow.

I am fortunate to have had many fine mentors throughout my life. They are too numerous to speak of individually, but I must thank all those folks who supported me and shared their advice and wisdom with me. I have never forgotten their kindness.

I am truly honored to receive this distinction. I thank the Career Communications Group and the selection committee for this recognition, and I thank all of you here tonight for sharing this very special evening with me.

Chapter 13

Paul Caldwell Jr.
1999 Black Engineer of the Year

THE FOLLOWING ARE EXCERPTS FROM THE 1999 BLACK ENGINEER OF THE YEAR AWARDS CEREMONY SPEECHES HONORING PAUL CALDWELL JR.

Mr. Eugene Renna, president and chief operating officer of Mobil Corporation: The Black Engineer of the Year Paul Caldwell came to be Chairman and General Manager of Mobil Producing Nigeria because he always—and I mean always—looks to make exciting things happen. The chairman of Mobil Producing Nigeria leads some 3,800 employees and contractors, producing 625 barrels of oil and gas a day with a budget of over a billion dollars. You don't get that job because you're a quota; you get that job because you're a producer.

Paul came to Mobil straight out of college and was promptly placed on a tank farm in Wilmington, California, tasked to clean up some old corroding equipment that had been decimated by chemical erosion. He set to work, took on the challenge, and made a noticeable difference. Another problem involved reducing sulphur dioxide emissions from the steam generators used to power refinery operations. Paul's team developed a very economic process for reducing these emissions. Paul was responsible for building the prototype flue-gas scrubbers and is one of the patent holders for this process.

As he developed experience and expertise in the oil-producing business, Paul kept alive a parallel track of personal development. He understood that in the world of business, seniority is not the principal criterion for upward mobility. So in 1979, Paul completed his M.B.A. at the University of Denver and afterwards became planning manager at Mobil Producing Netherlands where, among other things, he was responsible for negotiating commercial gas sales, including a $2-billion, 20-year deal with a European purchaser.

There were other intervening assignments and challenges at which Paul excelled. One that comes to mind is when he became vice president of Mobil Oil Indonesia. Here he managed oil and gas production and served as corporate liaison to the Indonesian government. Paul had production wells pumping out more than 50 million cubic feet of gas. As they were depleted, the standard practice was to drill more wells. One of his engineers suggested he drill the remaining wells with extra large tubing. Most wells have slightly less than three-inch tubing, but in Indonesia, Paul used nine-inch tubing. That was never done before. Some of those wells ended up with a capacity well over 200 million cubic feet of gas and will still be producing into the next century.

Well, you can see how Paul Caldwell came to be chairman and general manager, Mobil Producing Nigeria. The challenges of change are the steps he climbed to reach the top. But there is more to this story because, although this tells you what he has done, it does not explain the origin of his character. What got him here in the first place? I want to call upon the Old Storyteller to tell us about that.

The Old Storyteller: "Show me where a man is from, and I'll tell you who he is." Well, Paul Caldwell is from Oklahoma. And if you ask him where he was born, he'll say "on a farm, near Chandler."

See, Paul was born on a farm near Chandler, Oklahoma, only 40 years after that territory became a state. Up until 1907, Oklahoma was Indian Territory. It was the dusty destination for the "trail of tears."

The story goes that the government was moving the Indians from the fertile farming lands back East to this desolate, miserable dust bowl out West, where all the arable top soil had blown away in the winds that come sweeping down the plains. Later on, in the westward expansion, the government decided to open up this land for cultivation. They set a start time for the movement west to give everybody an equal shot at claiming their little parcel. Well, some folks couldn't wait to claim their little part of misery. They jumped the gun and ran ahead of everybody else. That's where Oklahomans got the name "sooners"— because they got there sooner than anybody else. But that is the backdrop for this man's character. Folks chasing challenges, rushing to claim their little piece of misery just to see what they could make of it.

So Oklahoma was where he was born, but on a farm, near Chandler, is where his character was carved. Now when you say "I live near a place," to city folks it means you're country. But to a country boy, it means you're special. It means you grew up with the values of the earth etched into your soul. On a farm near Chandler where Paul grew up, they didn't need any running water because their water came from a pristine well deep inside the bosom of the earth. They didn't want electricity lighting up areas of the room you didn't need; the kerosene lantern gave out just the right light for whatever you needed to focus on. And the challenges of an indoor toilet just don't measure up once you've had one outdoors.

Paul's mama was a school teacher in a one-room segregated school. But that little obstacle camouflaged a massive opportunity to focus his energies on the basic skills for a scientific future. I heard her say to him one day: "Do right, be fair, and manage your credibility. You are responsible for the image you leave folks with."

The whole family was engaged in teaching. His first six years of life, he lived with his grandmother. She taught him math by playing games and making him love it. Paul's father maintained the family farm and had a blacksmith business in Boley. Boley was an all-Black town founded by some industrious Black sooners who wanted to control their own destinies. His daddy taught him to live the spirit of Boley. He used to say, "Never bow down to any man who puts his britches on the same way you do." That lesson stuck. What Paul saw growing up was a world where everybody yearned to work. They never waited to work; they lusted after it.

His daddy also ran a scrap iron business. He'd go around the territory collecting discarded metal and bring it back and reshape and sell it. Paul was fascinated with everything that came back in that wagon. He couldn't wait to get his hands on the day's haul to see what it was made of. Oh, the pleasures you can get as a kid being able to dismantle the world with permission. This gave him a penetrating curiosity about how things work.

One story epitomizes the values that Paul accrued on the farm near Chandler that shaped his life forever. When he went to work at Mobil, he was taken out to a tank farm by Ted Bertness and shown miles and miles of corroding tanks and other petroleum paraphernalia in varying stages of disrepair. Paul said that Ted made the challenges come alive. Now an ordinary person, when shown endless miles of impending disaster, might be tempted to say, "Oh my God," and run the other way. But when Ted looked over at Paul, he had that sooner-stare—a look that said, "When can I get my hands on it to see what it is made of?" Paul Caldwell came to be a chairman at Mobil because he cultivated a passion for work on a farm near Chandler.

He and Mabel have got a daughter named Ime, which means patience. They want her to grow up to share their values, so they sent Ime to live for a while in a place certain to give her an adantage. You know where that is? On a farm, near Chandler.

Y'all have got to meet the chairman from a farm near Chandler.

1999 Black Engineer of the Year Paul Caldwell Jr.: Distinguished guests, ladies, and gentlemen. It is a great honor for me to accept this award tonight. I have a few people to thank, because no one gets anywhere alone. I want to thank my mother, who is no longer with us, because she taught me in school, at home, and throughout my life to stay focused on my goals. I thank my wife, Mabel, who supported me and made many sacrifices to go wherever the job took me.

I want to thank my first boss at Mobil Corporation, the late Ted Bertness, and a close associate, Dr. Earl Snavely, for welcoming me into a great company that gave me the opportunity to take on business challenges on a level playing field.

I thank my Mobil mentor, Kent Acord, for taking risks and believing in me enough to offer me jobs that built a career filled with as much challenge as reward.

I want to thank the man who nominated me for this award, Dr. Benne Bette, the manager of subsurface engineering at Mobil's research facility in Dallas, with whom I've had the privilege of working in the United States, Indonesia, and Nigeria.

And lastly, my thanks cannot be complete without recognizing Mobil Corporation, which provided the environment for me to grow professionally.

I want to share an experience that demonstrates how learning to solve business problems with technical innovation can boost your career.

When I was manager of producing technology in Dallas, the Zafiro field was discovered offshore Equatorial Guinea in West Africa. Kent Acord challenged my team to bring the field onstream 18 months after discovery. We tried an unusual procedure called "fast-track development." That meant conducting several activities simultaneously: we shot three-dimensional seismic studies, drilled appraisal wells, and designed production facilities at the same time. The project was onstream in 18 months instead of 40 and cost 30 percent less than expected.

I always involve the operations staff in a project early so they own it and feel part of its success. As you begin your own careers, you'll find the ability to work well with people is a crucial success factor you should nurture. It's a technique that transcends boundaries and one I have found effective living and working overseas. But inclusion goes beyond the work force.

In Indonesia, for example, not only was it important to get the most natural gas production possible from Mobil's giant Arun field, it was equally important to be good neighbors. I spent a lot of time with the community and the government to build and maintain good relationships.

The same goes for Nigeria. Nigeria is a nation of more than 100 million people from 150 ethnic groups with differing values. And Mobil Producing Nigeria is the second largest oil producer in that country. But to be a good corporate citizen, we work closely with our Nigerian partners and surrounding communities to make sure we not only produce oil, natural gas liquids, and condensate profitably, but also safely.

As good neighbors, we have helped build schools and hospitals, brought in electricity and clean drinking water to many small villages surrounding our facilities. We began educational training programs and continually hire local people to work at our facilities. It is an ongoing relationship that grows stronger every day. Some evidence of that is the number of our Nigerian friends who are here tonight. Thanks for your support.

If I were to offer advice to young people, I'd say don't be afraid to take risks. Aim high in your education. Focus your interest on what you like most. Know the business you're interested in. Treat all people with respect and dignity. Then soar as high in your career as your abilities will carry you. The world's best technology is still out there waiting for you to create it. Embrace the challenge. And step into the future with your mind wide open.

Thank you.

John Brooks Slaughter

1987 Black Engineer of the Year

Erroll Davis Jr.

1988 Black Engineer of the Year

Lt. Comdr. Donnie Cochran

1989 Black Engineer of the Year

Arlington W. Carter

1990 Black Engineer of the Year

Col. Guion Bluford

1991 Black Engineer of the Year

Arnold Stancell

1992 Black Engineer of the Year

Dr. James W. Mitchell

1993 Black Engineer of the Year

Dr. William Wiley

1994 Black Engineer of the Year

Walt Braithwaite

1995 Black Engineer of the Year

Lt. Gen. Albert J. Edmonds, USAF (Ret.)

1996 Black Engineer of the Year

Art Johnson

1997 Black Engineer of the Year

Lt. General Joe N. Ballard (Ret.)

1998 Black Engineer of the Year

Paul Caldwell Jr.

1999 Black Engineer of the Year

Dr. Mark E. Dean

2000 Black Engineer of the Year

Dr. Shirley Ann Jackson

2001 Black Engineer of the Year

Rodney O'Neal

2002 Black Engineer of the Year

Lydia W. Thomas

2003 Black Engineer of the Year

Anthony James

2004 Black Engineer of the Year

William D. Smith, P.E.

2005 Black Engineer of the Year

Linda Gooden

2006 Black Engineer of the Year

Part II:
The Second Decade

Chapter 14

Dr. Mark E. Dean
2000 Black Engineer of the Year

The following are excerpts from the 2000 Black Engineer of the Year Awards ceremony speeches honoring Dr. Mark E. Dean. The theme of the conference that year was "Education and The Leader's Imperative: Crossing the Digital Divide."

Dr. John Slaughter, member of the Board for IBM Corporation: The person we are called upon to honor tonight can rightly be called a renaissance man. If by renaissance man we mean a person who exhibits all the virtues of the innovative times in which we live, then surely Dr. Mark Dean qualifies, hands down. Mark received his Bachelor of Science degree in electrical engineering from the University of Tennessee in 1979. He came to IBM in 1980 straight out of school and immediately went to work on the team that developed the original IBM PC in Boca Raton, Florida. What came out of that collaboration was computer technology architecture that allows IBM and IBM-compatible PCs to run high-performance software and work in tandem with peripheral devices. This technology was first integrated in IBM PCs in 1984, and is now a key component for more than 40 million personal computers produced each year. Mark owns a share of three of the original nine IBM patents for that invention, and this was in his first years with the company. The PC AT [Advanced Technology], his next challenge, laid out the industry standard architecture for PCs. The AT was faster and could handle vastly greater amounts of data, and Mark Dean, the architect, wrote the standard. In 1982, Mark was awarded a master's in electrical engineering from Florida Atlantic University and a Ph.D. from Stanford in 1992.

Mr. Al Zollar, president and CEO of Lotus Corporation: After having been with IBM for 15 years, Mark was named an IBM Fellow in 1995, one of only 50 active fellows of IBM's 200,000 employees, the first African American to be so honored, and the 152nd person ever to receive the designation in IBM's 87-year history. Mark Dean holds more than 30 U.S. patents. He is currently director of the Austin Research Lab in Austin, Texas, where he just recently created the first chip powered at one gigahertz. That is one thousand megahertz. In 1989, in recognition of more than 20 patents issued or pending, Mark was inducted into the IBM Academy of Technology, and in 1997, he joined the Academy's high-profile Technology Council. Also in 1997, Mark won the Black Engineer of the Year President's award. He won the Ronald H. Brown American Innovators Award and was inducted into the National Inventors Hall of Fame. To date, only 137 men and women are honored in the National Inventor's Hall of Fame. Only two other Blacks, Dr. George Washington Carver and Dr. Percy Julian, have made it into the Hall. Last year, the National Society of Black Engineers selected Mark Dean to receive its top Golden Torch Award. During the year, he continued inventing things, reaching the tenth, eleventh, and twelfth levels in IBM's Master Inventor Award series. It takes at least three patents to bump up one level. *U.S. News & World Report* magazine recently reported on his latest dream, a wireless networked data tablet.

It has become a tradition to mark the extraordinary achievement of the Black Engineer of the Year with a remarkable adventure tale of how this life unfolded. So now I give you the Old Storyteller.

The Old Storyteller: Jefferson City, Tennessee is nestled down around the foothills of the Smokies, surrounded by pristine mountain streams and that autumn magnificence that only leaves from mountain nature can portray. It was here, cradled in heavenly splendor in the budding spring of 1957, that James and Barbara Dean made their mark on the world. Like all young couples just emerging from the enchantment of early marital bliss, they often talked of the legacy they would leave. And so in March, when their first-born arrived, they stepped out on faith, thought of their legacy, and named their first born Mark. Then with cunning tenderness and gentle persuasion, they sculpted their gift for the world.

The racial atmosphere in the South in the 1950s had spawned a difficult divide to breach. But James and Barbara were bridge builders. They understood that passage across any divide begins in the mind of the traveler. So they thought to construct a story-bridge for the safe passage of their Mark. Mark's daddy was a tinker, a man who loved to build. One time, he built a tractor from scratch while his eager-eyed prodigy son observed, enraptured by Daddy's work. The child thought, "he doesn't even see me watching." But what the child couldn't know was that the bliss of Daddy's work was a story pre-staged expressly for the audience of the child.

Mark's mother was a tenacious home-schooler. That means she made sure while Mark was at home, he was preparing for school. So she gave him math problems and watched with glee as he took to calculations like a tadpole in water. And every day as Mark exhibited greater facility with the numbers, there were these not-so-subtle, subliminal suggestions being etched upon his brain. He'd hear his mom whisper "you want to be an engineer." Not "You're going to be!" Not "You must be!" But "You want to be." She was giving over all the power to him.

In most schools, the child's environment is beyond the reach of the parents. But at Mark's school, Peck was Principal. Peck was only one of the most respected men in Jefferson City, Tennessee. But more importantly, Peck was Mark's mama's daddy. The esteem of his granddaddy was so huge, that even now when Mark goes home they say, "That's Peck's grandboy!"

Let me show you the effects of a story-bridge that's erected with cunning and care. In the first grade, Mark was tutoring his classmates in math. So advanced was he, that the teacher assigned this first grade wizard to fourth grade math. One day, Mark showed up at home with an algebra book. So his mom called a conference with the teacher and said, "I know he's bright, but what are you thinking, having him tutor upper grade kids in algebra?" The teacher said, "Why not? He does it at school." His original teacher and biggest fan was shocked at the pace of his learning.

This accounts for the brilliance, but what of the character? How does a child develop the tenaciousness of static cling, or the confidence of a prince, to just plain out overwhelm obstacles? At Mark's house was an ancient soul, the matriarch and keeper of the ancestral wisdom. She was his great-grandma who died nine years ago at the age of 107. She completes the story-bridge that forms a passage across the divide. For it was she who told him stories of when times were really hard. His grandma told him some horror stories and some precious gems, which I can't go into here, but she always ended her story with this little caveat: "Times were hard but they can get better. We just got to keep at it." Mark thinks even now, that up against what she had to endure, his little problems pale in significance. So he just keeps at it. Not perfection, just good enough is his credo.

I can go on all night about this bright and shining star. But you see, I don't have to. I can capture his life in one culminating legend about the effects of his parents' story-bridge. When grown up Mark was going for his Ph.D., he noticed that the program called for spending six to seven years in study. He said, "That's unacceptable. I'll do it in half the time." And he did just that.

How do you cross a digital divide? Well, some 44 years ago, some remarkable people set out to construct a story-bridge to set their mark upon the world. And what a mark they've made: The Mark of the Dean star. And with one so young, so bright, and carrying centurion genes from his great grandma, you ain't seen nothing yet. This star is a comet and we're seeing only the brilliance of the tail.

2000 Black Engineer of the Year Dr. Mark Dean: How can a country boy from a small town in the Smoky Mountains of Tennessee be standing on this stage to receive such high recognition from people he respects?worthy peers, capable colleagues, and industry leaders who have their own remarkable talents and achievements to boast? It is because this award represents the contributions and sacrifices of multitudes who guided me in the discovery of my potential. So tonight, I stand here not as a winner but as a presenter—to those who did the real work behind my success.

Tonight this award recognizes a teacher who, despite struggling with four grades in a single classroom, took the time to encourage one first grade student to complete fourth grade math, much to the initial shock and eventual delight of his parents.

It goes to wonderful parents and grandparents who planted the seeds of determination, drive, and a "can-do" attitude to ensure that few limits existed for a son's ability to achieve.

This recognition represents the success of minority engineering scholarships and college diversity programs, which opened a pathway for a young man to experience the exhilarating challenges and raw excitement of engineering, and to graduate at the top of his form and his class.

It recognizes the support and sacrifice of a spouse who helped her closest companion complete his Ph.D., and who continues to encourage those weird and unusual ideas.

With this award I highlight the efforts of a company to provide opportunities to minorities, asking only in return for hard work and a perspective unavailable in a mono-colored, mono-cultured environment.

Enlightened leaders in this company gave an eager young engineer unparalleled chances to develop technologies, which established the standards for today's personal computers; drive a team producing the world's first 1,000-megahertz microprocessor; and rigorously investigate technological requirements that will make the Information Age accessible to everybody, independent of race, age, or income.

Although this honor is bestowed upon an individual, it recognizes the guidance, selflessness, and support of the unsung heroes who made it possible for this individual to achieve. Yet all they required of me was to accept no barriers, pledge to work hard, believe in myself, and strive to enjoy life. What a magnificent formula.

It was these mentors, helpers, guides, partners, supporters, and cheerleaders who taught me that everybody has great potential to make a grand difference. And when we accept that opportunities are constrained merely by the limits of our mind, only then can we help young people to uncover and leverage the wild and lucrative possibilities of the Digital Revolution.

I sincerely appreciate the opportunities you gave me, allowing me to make a difference. This award means a lot to me and I hope it means as much to you who truly made it possible.

Thank You.

Chapter 15

Dr. Shirley Ann Jackson
2001 Black Engineer of the Year

THE FOLLOWING ARE EXCERPTS FROM THE 2001 BLACK ENGINEER OF THE YEAR AWARDS CEREMONY SPEECHES HONORING DR. SHIRLEY ANN JACKSON. THE THEME OF THE CONFERENCE THAT YEAR WAS "HERITAGE BY CHANCE, SUCCESS BY CHOICE."

Mr. Robert Stevens, president and chief operating officer of Lockheed Martin: Isn't it curious that the chance of a lifetime can be totally obliterated by the choice of a single moment? Conversely, one momentous choice can turn the most insignificant event into the chance of a lifetime. Choice giveth and choice taketh away. In the past 15 years, this conference and celebration have done much to strengthen the choices youngsters make regarding their futures. The reign of the Black Engineer of the Year illuminates the message to the young that where others have gone, there may you go also. This is the first Black Engineer celebration of the new millennium, and so our charge to the Black Engineer of the Year is simple yet profound: may your brilliant light shine so bright that it penetrates all the dark corners of the world, where ignorance and apathy abound, and lead all our children home to the fountain of knowledge. Dr. Jackson, Lockheed Martin is proud to team up once again with the engineering deans of the Historically Black Colleges and Universities in sponsoring this conference and to welcome you into the fold and to wish you well.

Nicholas Donofrio, senior vice president for Manufacturing and Technology at IBM: By education, training, and profession, Dr. Shirley Ann Jackson is a theoretical physicist.

Rodney Adkins, general manager of IBM's WEB Server Group: By experience, position, and calling, Dr. Jackson is a genuine leader.

Mr. Donofrio: As a physicist, she sits on the board of trustees of her alma mater, MIT, from which she holds the distinction of being the first African-American woman to earn a Ph.D. in any discipline.

Mr. Adkins: As a leader, she sits at the helm as president and CEO of Rensselaer Polytechnic Institute. It is the oldest technological university in the United States, at which, as president, she holds the distinction of being the first African American to hold that position.

Mr. Donofrio: Professional scholarship led to Dr. Jackson's appointment as the first African American to serve on the Nuclear Regulatory Commission.

Mr. Adkins: Managerial leadership led to her becoming the first African-American woman appointed by President Clinton to chair the Nuclear Regulatory Commission.

Mr. Donofrio: Dr. Jackson's scholarship sparked her contribution to AT&T Bell Laboratories when, for 15 years, she performed explorations into solid-state physics that led to the improvements in signal-handling capabilities of semiconductor devices. In other words, she contributed to Bell Labs' lead in electronic communications. In addition to MIT, she studied in Sicily and France, then lectured at the NATO Advanced Study Institute in Antwerp, Belgium, the Stanford Linear Accelerator in California, and was visiting scientist at the Aspen Center for Physics in Aspen, Colorado.

Mr. Adkins: In addition to serving on the MIT board since 1975, Dr. Jackson's leadership has led her to serve on the board of Lincoln University, Rutgers University's Board of Trustees and Board of Governors, the board of Associated Universities, which operates the Brookhaven National Laboratory, and the National Radio Astronomy Observatory. In addition, Shirley joined the board of the Brookings Institution in June 2000.

Mr. Donofrio: As a member of the National Research Council's Committee on the Education and Employment of Women in Science and Engineering, Dr. Jackson diligently promoted the advancement of women in science.

Mr. Adkins: Promoting greater opportunities for women is leadership.

Mr. Donofrio: Promoting the advancement of women in science is supporting scholarship.

Mr. Adkins: Leadership!

Mr. Donofrio: Scholarship!

Mr. Adkins: Well, here is something we can agree on: Shirley Jackson has continued to teach and publish throughout her outstanding career and in 1998, was inducted into the National Women's Hall of Fame for her Leadership contributions.

Mr. Donofrio: And just last year, Dr. Jackson won the National Society of Black Engineers' Golden Torch Award for Lifetime Achievement in Academia. And we both know that leadership is to scholarship what telling is to teaching. You can't truly have one without the other.

Mr. Adkins: Dr. Jackson's career is a remarkable example of teaching. Now let's have an example of how she came to be a teller. It is a tradition for the Old storyteller to offer a short retrospective on the life of our winner. Here now is the Old Storyteller.

The Old Storyteller: George and Beatrice Jackson were by any measure magnificent parents. It isn't clear where they learned to become such exquisite sculptors of human excellence, but the whole neighborhood of Southwest Washington, D.C., understood that if you wanted magnificence in children, you go to George and Bea. I heard them say once, "You don't make kids become somebody, you help them. You cultivate their spirit, be seen to respect their interests, and always, always, always, nourish their natural zeal. Then the instant you see the flame take hold, become a model for their values and kindly step out of their way." Neither George nor Bea was ever a physicist or university president, not a flourishing attorney nor a vital college dean. But they reared three girls who are.

But I know how they did it cause I was there, and I'm here tonight to reveal their secrets. You see, over each newborn child, George and Bea would linger for a while and wonder aloud, "What great surprise do you bring into our lives, little one? What is your fate, your special gift, and how can we help you bring it forth to flourish?" Then they would put the baby down to begin the long, loving, parental watch for the special sign the child possessed to illuminate her purpose.

When Shirley Ann, the second of their four babies born, was coming of age, George and Beatrice saw their sign, for a vigilant restlessness always engulfed her. She was curious beyond measure, this child, about everything. But Shirley Ann's curiosity was unusual. She wasn't just childishly inquisitive; she was a thorough investigator. Whatever it was, this child just had to touch it, turn it, or twist it, then capture or cultivate and decapitate it; she kept a log of her explorations. "But," George said to Bea, "she then deeply studies her catch with the eye of a trained detective, and when she is satisfied, there is no more to be learned, she puts it all back together again. Or, praise be to heaven, she tries."

Shirley Jackson's life seems to have been shaped in the mold of the scientific method. Now George himself, though largely unschooled, was a human calculator and a wizard with his hands. Beatrice was the family theoretician. She had schooling and had taught for a while, so she knew how to channel curiosity into action. With these two for parents, this prodigy of a child could not have been born to greater advantage. So when Shirley Ann became of age, the whole world felt the tremor. In school, she was perhaps the only child ever put out of the reading circle because she read too fast. But even when she was put out, she would read her own books silently across the room and listen to the club with that portion of her magnificent brain that was unoccupied at the time, and call out the answer to questions from books they were reading. How do you raise a genius? You teach them good values, then get out of their way and watch them soar.

When she was a little girl, the children use to call her Shirl the Squirrel. Now, children are remarkably clairvoyant, able to look deep into the soul to reveal the fundamental nature of things. If you listen closely to what children call things, they'll precisely pinpoint its natural imperative, then package their findings in an affectionately contrived nickname with which to taunt the unsuspecting bearer. Shirl was called squirrel because of her relentless pursuit of excellence. But why squirrel? Far beyond the pedestrian notion that it happens to rhyme with Shirl, there is a deeper, more prophetic meaning to this soubriquet. Andy Longclaw, a Native American manager, believed that the first secret to motivating people is to encourage the spirit of the squirrel. The squirrel, Andy's grandfather told him, works with passionate zeal because she understands the true value of her work, how her every effort connects directly to the salvation of her world.

At MIT Shirl was called the computer. Now, the minimum basic entrance requirement at MIT is that you be a veritable wizard. The last-place graduate from the lowest performing year at MIT is still thought of as genius. So when a cohort of geniuses nicknames you "the computer," not much else need be said.

It is fitting that IBM submitted this nomination of Shirley Jackson, for they are the repository of Big Blue—that other super computer?so they must know one when they see one. It is also fitting that George Jackson always told this child, as you will hear her say, that she should climb to the top of trees. For then you will understand why I named this story, "The Legend of the Big Blue Squirrel."

2001 Black Engineer of the Year Dr. Shirley Ann Jackson: My father was a U.S. Postal supervisor and my mother a caseworker in the Department of Human Services in Washington, D.C., where I grew up. Between them, I was well groomed to seek the best, the most, and the highest of whatever I attempted. When I was growing up, we did a lot of projects and learning together as a family. My parents helped us design and build go-carts and cages for mice that I used in nutrition experiments. We even designed a slide rule based on Boolean algebra. In the summer, we attended special programs, did household chores, swam, and read together. From 10 years old to high school, I spent my summers collecting live yellow jackets, bumble bees, and wasps. I manipulated their diet and environment, as I observed how it influenced their behavior and vitality. Some didn't survive the vitality test. But my father always said, "Leap for the stars, so that you can reach the treetops. You may not hit your target, but at least you'll get off the ground."

This advice has helped me throughout my life. It was most especially helpful when I arrived at MIT in 1964. It was the end of "Freedom Summer," a long, hot summer when racial tensions were running at fever pitch. On the campus of MIT, there were less than a dozen African-American undergraduate students when I arrived, and I was instantly isolated. Study groups would convene and routinely leave me out. After I got my tenth "A" on exams, they realized what I could do, that I was as serious and committed as they. In a word, that I was good. They began to change. It was not a sudden change. Initially, people only wanted to talk about a specific problem they were having, but eventually, and somewhat begrudgingly, I was allowed in. Although it took a while, some of them became my friends.

Isolation can be a serious defeat to even the most committed of academic ambitions if one is unaware of the value of their chances or unarmed as to the abundance of their choices. My father's admonition to leap for the stars let me know that the academy was the best place to equalize my chances with stern and unalterable choices. And so, for me, defeat was not an option.

Sometimes a veil of discrimination comes in the form of discouraging words. While I was still deciding on a major at MIT, a professor offered me a bit of career advice. "Colored girls," he said, "should learn a trade." I pondered this advice and again thought of the choices available to me. Chance had made me colored, I was by chance a girl, and I was here to learn. And as for the trade? Well, I chose physics. And I have been trading very well in that domain for these many years.

Discouraging words can deflate career ambitions only when those ambitions are not anchored in rock-solid principles. Mine were—thanks to my family—and I was able to overcome.

Chance is a two-faced chameleon. She always appears as both opportunity and obstacle. The anatomy of choice is the ability to determine your fate by the selections you make. When I was faced with the options to give in to ignorance or go on to excellence, I chose to persevere.

From the moment we learn to select from among the many chances we encounter, our lives become the product of our choices. Doing nothing is as much a choice as is doing it all. Given that, why not leap for the stars? You may never catch one, but at least you'll be off the ground.

I'm grateful to my father for his sage advice. I want to thank Ms. Marie Moss Smith who has inspired me along the way.

I'm grateful to Ted Childs at IBM who nominated me, and to Tyrone Taborn and the Career Communications Group and the engineering deans for having had the courage and foresight to shine the light on those who have been reaching for stars without recognition for centuries.

And thank you for this delightful reception. May all your chances be blessed and all your choices be wise.

Chapter 16

Rodney O'Neal
2002 Black Engineer of the Year

The following are excerpts from the 2002 Black Engineer of the Year Awards ceremony speeches honoring Rodney O'Neal.

Mr. Robert Stevens, President & Chief Operating Officer of Lockheed Martin Corporation: Rodney O'Neal is an executive vice president of Delphi Automotive Systems and president of the company's Safety, Thermal, and Electrical Architecture Sector. He is a member of the Delphi Strategy Board, the company's top policy-making group, and serves as executive champion for Delphi's Ford customer team. Now, you may not know Rodney or Delphi Automotive, but I assure you they've both touched your life in some way. Let me tell you why.

Delphi Automotive Systems is a world leader in making components for your car. They have three business sectors with 195,000 employees operating in 43 countries. They make components like window defoggers, climate control systems, electrical wiring designs controlling ignition, lighting and sound systems, and other parts that are at the heart and soul of the safety, comfort, and convenience of your car. In fact, Delphi Automotive is the largest producer of automobile components in the world. They are even under contract to supply the electronics for the Segway Human Transporter, destined to scoot itself into our lives in the near future.

Rodney O'Neal is one of the top five corporate officers leading Delphi Automotive and the chief executive of the sector directly responsible for electrical systems, heating and air conditioning components, and creative solutions for automobile safety. He is responsible for 110,000 employees.

Rodney began his automotive career as a student at General Motors Institute in 1971. In 1976, he joined the Inland Division, where he held a number of engineering and manufacturing positions in Dayton, Ohio, Portugal, and Canada. Rodney was named director of industrial engineering for the former Chevrolet-Pontiac-GM of Canada Group in 1991, and the following year became a director of manufacturing for the former Automotive Components Group Worldwide, now Delphi Automotive Systems. After successive promotions to positions of increasing responsibility between 1994 and 1998, Rodney was named to his current position in January 2000.

To help me introduce you to this year's Black Engineer of the Year, I give you now the Old Storyteller, who will tell us how Rodney O'Neal came to roost at the top of this magnificent perch.

The Old Storyteller: I remember when Rodney O'Neal was just a little Dayton daydreamer. He would sit in the classroom most of the day, staring out the window, lost in a dream. Ole Miss Miller, his teacher, would call his mother, Ida B., and complain about his daydreaming. Then when James, his dad, came home from work for dinner, Ida B. would say, "James you gotta talk to Rod about the daydreaming." James would say, "Well now Ida B., you know you can't advise the inadvisable. He's just a free spirit. And you can't capture a free spirit. The only way to control one is to lay a solid foundation, set a sterling example, and hope and pray that it will catch you. So they set out to cement an indestructible foundation in which to plant their chances.

They decided that a strong family would give them a huge advantage. James made sure his three boys saw that there was nothing you can't do. "Can't" was just a coward's argument for not trying.

Rod grew up watching his parents gardening, rebuilding engines and fixing any other mechanical contraption hanging around the house. He even saw his daddy build a house with his own hands.

That was huge. But Rod never heard them utter a single word of pity. No sir, he grew up hearing his parents saying, "Education will give you chances you can't get no other way," or, "All you have is your word, and if that ever degrades, you are a wretched being." Some days, the family drove around looking at dream houses. Rod would say, "Wow, they must make a lot of money to own a house like that." James would whisper, "You know it's not what you make—it's what you keep."

Pitch, patch, a free spirit catch.

Percy Jones, Rod's high school math teacher, was a genius, but one day he told Rod's parents, "This boy daydreams too much. He can't cut it in math. If I were you, I'd think about taking him out of the class." Rod agreed! But his parents had a genius for guiding free spirits. They understood that daydreaming signified a vigorous mind. Their boy was not just smart, he was brilliant. But he would never tolerate being told he had to stay in that class. So James said to him one day, "You know, we're going to take you out of that math class because that's what you want. But that teacher doesn't believe you can cut it. Now, if it was me, I would stay in there just to prove him wrong. But since you want out, we'll take you out."

Bing! Bare! A free spirit snare! Rod stayed and showed old Percy Jones that he might be a genius in math, but he didn't have a clue about free spirits.

Well, in 1970, Rod was looking for a college to study computers like his brother. Money was scarce, so his search was going slow. Now, it happened that at that very same instant, General Motors was seeking minorities to attend General Motors Institute. The applicants had to be exceptional on grades, test scores, and deportment. But just as Rod was looking for a college, Mrs. Richardson, the guidance counselor, was looking for him—and only him! She said, "Rodney, you should apply to GMI." But he'd never heard of GMI, so he promptly forgot about it.

But Mrs. Richardson had some experience with free spirits herself. She filled out the application, coaxed him to sign it, and then explained at GMI, there were no girls and no sports, but they paid you to study. A zip and a zap, and a free spirit trap.

Here is the sum of it all. Everybody he met could have left Rodney to his own devices, but they didn't. Why? Because what they saw in him was far, far greater than what they didn't see. They saw what his long-time manager said to him as he was moving up the corporate ladder to success. He said, "If you climb on our shoulders, we'll take you up to the top. All you have to remember is to never dig in your heels."

Well, that's all I've got time to say right now. But whenever you see a child just lost in a trance, don't be so quick to write him off. It may be a free spirit, daydreaming his way from the bottom of Dayton to the top of Delphi. And if you can't imagine that, just remember the Black Engineer of the Year for twenty-oh-two, Rodney O'Neal!

2002 Black Engineer of the Year Rodney O'Neal: It's truly an honor to be here this evening to accept this award. It's also, frankly, a little humbling. If you had asked me 35 years ago, as a young teenager growing up in Dayton, Ohio, to define success, I would have told you that the world would be mine, if only I could make $50,000 a year, have a nice crib, and a slick set of wheels! That was my dream. It's all I could imagine.

I certainly would not have imagined this evening. I would not have imagined opportunities to live and work all over the world, handling large and complex corporate operations. I would not have imagined knowing other African Americans in similar positions. I would not have imagined these things because, as you know, back then, America was a different place. It was a place where my father—a creative, technically gifted man—could not get a mortgage. It was a place where there were no examples of African-American executive role models. It was a place where the halls of power—and for that matter, many restaurants and public restrooms—were closed to me.

So, when people ask me, "Did you have a master plan for your success?" I can only give one answer: no, because I had no idea that this sort of achievement was even possible, even in my dreams.

I had no master plan. What I did have from the start were people who believed in me. I had my dad who told me how far I could go, who made me believe I had a gift for technical things, who helped me see my own leadership abilities, and who saw my free spirit as an asset—even when others just saw me as a daydreamer.

I had a guidance counselor, Mrs. Richardson, who knew that without financing there would be no college for me, and who wouldn't let up until I sent in my application to General Motors Institute, now called Kettering University.

I had a mentor, Bill Quilty, who talked me out of dropping out from GMI and quitting General Motors, probably a million times. No matter how much I sometimes struggled with the vastly different culture, the loneliness, the sense of isolation as an African American, Bill managed to keep me there, advising me to stay and change things instead of quitting.

It has always been my good fortune to have been surrounded and supported by individuals who cared about me, people who unselfishly lifted me on their shoulders and carried me to success across the abyss of failure.

So, I dedicate this award to all the people who made it possible for me to be here: The African-American trailblazers of the past, my friends, and all of the wonderful individuals who took the time and made the effort to help me reach my full potential.

Most importantly, I want to thank my family who supported me, loved me, and reminded me of what was really important, over and over again. My pillar of strength, my lovely wife, Pamela. My princess, my daughter, Heather. And the baby boy, my son, Damien. You truly are the air underneath my wings.

My metrics for success have radically changed since those days back in Dayton. I've come to gauge my success by the number of hearts I've gathered during my lifetime. And, by that account, I am one of the most successful and richest men in the world.

For me, this award represents the power that flows from the love and help of others, and it is indeed a great honor.

Finally, to all the young people out there: dream big. Imagine the impossible; imagine the improbable. It can come true for you. And above all, please remember that to get anywhere in life, you have to do it yourself, but you can't do it alone.

Thank you again.

Chapter 17

Lydia W. Thomas
2003 Black Engineer of the Year

The following are excerpts from the 2003 Black Engineer of the Year Awards ceremony speeches honoring Lydia W. Thomas. The theme of the conference that year was "Stretch for the Breakthrough, Soar Beyond the Limits!"

Robert Stevens, president and chief operating officer of Lockheed Martin Corporation: Dr. Lydia Thomas has built an extraordinary career based on one simple concept: curiosity about the world we live in. Dr. Thomas has an unquenchable thirst for knowledge about the Earth and its infinite variety of inhabitants. This is what led her to earn her bachelor's degree in zoology, her master's degree in microbiology, and her Ph.D. in cytology, which is the study of cells.

It is also what led her to join The MITRE Corporation, where she specialized in the people sciences and environmental issues. As but one example of her many accomplishments in this arena, she led the effort to select the Superfund sites, areas of our country targeted for a much needed clean-up of toxic waste. Her talent and intellect were immediately recognized at MITRE, and after a rapid rise through the management ranks, Dr. Thomas was tapped to lead a spin-off company called Mitretek Systems.

As President and CEO since its inception in 1996, Dr. Thomas has lived the company's motto, providing innovative technology in the public interest. Now employing more than 700 professionals—60 percent of whom are scientists and engineers with advanced degrees—Mitretek has become a leader in providing research and engineering capabilities that benefit people. Right now, for instance, the company is fully engaged in the application of new and emerging technologies in the areas of criminal justice, health care, environmental and energy sciences, counter-terrorism, toxicology, and homeland security—just to name a few.

As we are well aware, these technologies are tremendously important to all of us right now. And Dr. Thomas's stewardship of that effort is extremely reassuring. Among an impressive list of firsts and select appointments, I unfortunately have time to mention only a few. In 1991, Dr. Thomas was honored by this forum with the prestigious Dean's Award. More recently, she was appointed by the President to serve on the Homeland Security Advisory Council and was elected to the Council on Foreign Relations. She serves her company, she serves her community, and she serves her country with equal energy and commitment.

Clearly, Lydia Thomas is a super-achiever. What may be less clear is exactly how she got that way.

2003 Black Engineer of the Year Lydia W. Thomas: What a terrific evening. I am pleased, grateful, and humbled by this wonderful recognition.

When I was growing up, my high school principal was well known by an aphorism he frequently used. No matter the problems, our principal would say, "Only your best is sufficient." Now, my principal was also my father, which gave me a unique view into the passion behind those words and served to burn them deeper into my character. Tonight I'm reminded of my father as I witness with you the best in all of these winners. I congratulate you. As for me, I hope my father would have agreed that I've done my best and in the end, it will be sufficient.

Growing up as the principal's daughter was not always easy. But then, one doesn't have to stretch to get beyond easy, and stretching yourself to set, and then exceed, the highest expectations you set for yourself was what my father instilled in me. My father was a word collector who used his newfound words in conversations. I often had to get the dictionary to understand him, but it gave me an advantage that I yet cherish. Now, my father was also the smartest man I knew, and I could never find an error in his logic. It was frustrating. But when I got to Howard and began to study science, I happened upon a question that stumped him cold. Of course, I bragged. That was a mistake. While I was gloating from my victory, my father wrote me a letter, in English, filled from start to stop with indecipherable words. It took me weeks to read it. When we spoke on the phone he simply asked, "Did you get my letter?" and we never mentioned it again. You see, for my father, retribution was to stretch my mind and lead me up to heights where I could soar. So tonight's theme is very reminiscent of a glorious time for me.

I have been very fortunate in my life. The people who wanted me to soar have always far outnumbered those who would clip my wings. It was they who taught me to always be aware that there is no force conceivable that can long endure what you put your heart and mind into achieving. So while my gratitude is overflowing, to thank them all would take all night.

So indulge me as I thank my teachers, from K through Ph.D., who built the foundation of education upon which I stand.

My colleagues at Mitretek and MITRE, thanks for teaching me to put my skills to work for the benefit of society.

My beautiful and brilliant daughter, who put up with a too often absent mother, thank you for your patience and support.

And the rest of my wonderfully supportive family and friends. All of you, along with my adoring parents, made this night possible for me; it is as much yours as it is mine.

Finally, I am grateful to you, Tyrone, and your stalwart staff of warriors who often fight through gales to make these recognitions happen. The deans and those of you who sponsor these events, it is highly laudable and deeply appreciated by us all.

If you set your expectations high and keep stretching for the breakthroughs, you will certainly come to soar well beyond all the limits.

Thank you!

Chapter 18

Anthony James
2004 Black Engineer of the Year

THE FOLLOWING ARE EXCERPTS FROM THE 2004 BLACK ENGINEER OF THE YEAR AWARDS CEREMONY SPEECHES HONORING ANTHONY JAMES. THE THEME OF THE CONFERENCE THAT YEAR WAS "50 YEARS AFTER BROWN."

Allen Franklin, chairman, president, and chief executive officer of Southern Company: How do you become an electric utility CEO for one of the largest power companies in the world—an electric utility with a service area of 2,000 square miles of prime southern real estate, revenue of more than $300 million, and 137,000 customers? Well, if you are born with the drive and determination of Anthony James, it's just a matter of time. Let's begin at the beginning. In 1973, Procter & Gamble hired a young engineer from the University of South Florida to work in its paper-products plant. Within five years, he advanced through six managerial positions, winning promotions faster than anyone in the history of his division. When he moved to Southern Company in 1978, Anthony again began a steady rise through the ranks. He progressed from supervisory positions to executive management by demonstrating an eagerness to think strategically and analytically, be proactive, work hard, and generally make the entire team better. Anthony's combination of engineering, managerial, and people skills was evident when he simultaneously managed five Georgia Power generating plants, one of which was the largest coal-fired power plant in America, while also serving as vice president of power generation for Savannah Electric.

In 2001, Anthony became president and CEO of Savannah Electric. And the city of Savannah, Georgia, welcomed its first African-American corporate CEO with proud, open arms. As CEO, Anthony continues to be a mentor for young minority engineers and professionals, both inside and outside our company. He strives to give back to his community what others gave him growing up. Anthony James is an engineer, a communicator, a leader, a darn good guy, and one of Southern Company's most valuable assets. Now, to understand how Anthony put himself in a position to be all these things, let me present to you the old storyteller.

The Old Storyteller: This is the legend of Lake Alfred and the Saints. It was a hot Florida summer in 1950 when Neris and Deforest James prepared to name their second child. Neris was deeply religious and played piano in the choir for the church next door, so whatever this baby would be named, it would have something to do with the power of belief. Now, I don't know if you know this, but all parents look for names that will give their baby the greatest advantage in life. Neris and Deforest were no different. The family name was James, and they were aware of St. James the apostle. So when Neris and Deforest came to name their baby, they named him after Saint Anthony to give him the power of two saints. A saint is a person who has been canonized. Canonized means that he lived his life in such a way that changed the rules, raised the standard, moved the bar for everybody else. So Neris and Deforest understood that, and set out to shape their baby's character with his name. They knew that the characters that spelled the name were not as important as the character behind the name. So they set out early on to inscribe on this lad's character the way of the saints.

When Anthony was young, Deforest took him out to the fields to work. Sometimes they picked oranges, loaded watermelons, dug holes for USDA, so that when he reached teen age, he could work by himself. Young Anthony was hired one time to dig for nematodes. Now, a nematode is a parasitic worm that feeds on the roots of orange trees. If not caught, he'll devastate a grove. So one day, Anthony went to dig for Nematodes and promptly learned two crucial lessons for life, one from the nematode and one from the earth. The nematode taught him that to get anywhere in life, you always cling to the roots because there is where you'll find your greatest sustenance. The earth taught him that digging for nematodes is back-breaking work and not something you want to do for a living. Well now, he knew what direction he didn't want to take, so he just needed directions on what he did.

When Anthony went to high school, he found math and science to be easy. Now, that was hard to explain because he was coming from across the tracks and wasn't supposed to be as smart. But what he didn't know was that all the time Deforest was taking on those jobs, he was chiseling a message in the young boy's character that could not be removed, no matter what kind of eraser you used. See, Forest, as his dad was called, was the neighborhood fix-it man. If it could break, Forest could fix it. So while he worked with the passion of a saint, young Anthony was in the direct line of fire. So when Anthony got to college, he met Andrew Minor. Andrew—another Saint. He was telling Andrew one day that he was lost for a career and Andrew said, "Listen, NASA has got a co-op going on. Why don't you see if you can sign up?" Well he did and went over there and got mixed up in Doppler Radar designs and space shuttle assembly and then he looked up at the stars. What most people saw when they looked up there were stars. But Anthony looked up to the stars, then past the stars, and then beyond the stars, and no where could he see a ceiling. The sky was the ultimate opportunity and it had no cap. He saw the sky as parallel to life and it had no limits. Probably nobody else ever had such as experience.

It had to be here that all those stories he had ever heard came together. The canon of life, the rules that you live by are all set by how you interpret the meaning of your experience. So although he carried the name of two saints, the ones who set the canon were those who told the story. And he just took those stories and lived by them. So if you go to Lake Alfred in search of the saints, you might look for them in the fact that his grandmother Mary, who taught him business, had a sister named Martha. You might look at NASA and the Nematodes as giving him guidance. You might think about the two names of saints he was given. But if I had to bet, he would probably say the saints that gave him his canon, the ones who set the standard for him to raise, would be Neris and Deforest, because they are the reason that we are able to say that the Black Engineer of the Year for 2004 is Mr. Anthony James.

2004 Black Engineer of the Year Anthony James: On behalf of all the award winners tonight, I want to thank Career Communications Group and all the sponsors who have made this wonderful celebration possible. It has been more than I could have ever imagined and a night that I will never forget. And thank you, Allen, for being here to share this special moment with me. My career has been built on the support I have received from so many Southern Company employees, and my appreciation extends to the entire Southern Company and Savannah Electric family. I want to thank my wife, Sheila, who was supporting me before I even started dreaming of being an engineer, and our son, Tony, who has always been an inspiration for me to be all I can be. I'd also like to express my appreciation to my sister, Yvonne, my nieces, and all the other relatives and friends who have journeyed here to be with me tonight. My mother, Neris James, was unable to be with us tonight, but I want to thank her for all the incredible support she gave me growing up and for one decision she made in particular. When I was 16 years old, my father was working a construction job, helping to build Disney World. On his way to work early one morning, a car crashed into his car and he lost one of his arms. I knew immediately what I needed to do. I went to my mother, and I said to her, "I'll go to work." She looked right back at me and said, "No, you're not going to drop out of school. We can make it." So I stayed in school and found two or three part-time jobs. My mother always encouraged me to be the best I could be in whatever I did, and I have always tried to follow her advice. But I never thought it would lead me to this type of recognition. It far exceeds anything I thought I could achieve in my lifetime and is proof that if we persevere and work hard, who knows what we can achieve over a 30- to 40-year timetable. It is also proof that when we allow ourselves to dream and give it our best shot, good things can happen. The poet Langston Hughes said, "Hold fast to dreams, for if dreams die, life is a broken-winged bird that cannot fly." I would tell you not only to keep dreaming but also to help others achieve their dreams. Go into the schools, talk to students, become a mentor. We are all standing on shoulders of giants, people who have gone before us and laid the path that we now travel. Not one of us can go it alone. I congratulate tonight's 19 other award winners. And on their behalf, I say thank you to all of you who have made our dreams possible.

Achievement: Tonight we have celebrated the story of an overcoming. It is a tale fairly acquitted, barely admitted, and nearly forfeited. Yet in the annals of all humankind, there exists no greater legend.

The Old Storyteller: It is the story of the Black experience that has thoroughly replaced a defaming fable of failure with a transcending saga of success. In the entire history of the known world, no people has ever achieved the virtually complete and remarkable overhaul of their story in the face of its continuation.

Achievement: To take a history so thoroughly replete with failure, impotence, and negativity, and supplant it as it's being told with positive tales of achievement, power, and pursuit is a feat worthy of the grandest celebration, and we are come to make it right.

The Old Storyteller: So we have learned that while the elements of success are fundamental, they can only be accessed through a story. Where there is no story of access, there can be no desire for success. For desire is based almost entirely on possibility, and possibility for the future is connected to the legends of the past.

Chapter 19

William D. Smith, P.E.
2005 Black Engineer of the Year

THE FOLLOWING ARE EXCERPTS FROM THE 2005 BLACK ENGINEER OF THE YEAR AWARDS CEREMONY SPEECHES HONORING WILLIAM D. SMITH, P.E. THE THEME OF THE CONFERENCE THAT YEAR WAS "SEIZE THE EXCELLENCE IMPERATIVE."

The Honorable Rodney Slater, former secretary of United States Department of Transportation and partner with Patton Boggs LLP: It is truly an honor for me to be here tonight to introduce you to my friend Bill Smith. It is a double honor to do so while presenting him with the top award that is connected to the word "excellence." For that is precisely how I see Bill's career. Allow me to offer just a few highlights.

Bill's road to excellence began at North Carolina A&T State University, one of the registered Historically Black Colleges and Universities, training to become an electrical engineer, an ROTC Air Force officer, and interning at IBM. Upon graduation and redeeming his military obligation, Bill apprenticed at firms like Kaiser Aluminum and Bechtel Corporation in a variety of diverse assignments: mining projects in Indonesia; copper refining in the American Southwest; iron mining in Brazil, ore crushing in Chile; and rock quarry operations in California. It seems that from the start, Bill's climb up to the top began by digging down to the bottom of things. He was identified for leadership early on, and appointed to the Directors' Advisory Group at Bechtel Corp., a select group of senior manager candidates who advised top executives on strategic issues. In leadership positions around the world, Bill was often the first Black American engineer some of his overseas mining colleagues ever saw. But, he mastered their language and culture to motivate them in their language.

Bill came to Parsons Brinckerhoff in 1988 as chief engineer of the FMC Power Group. The CEO of PB became his mentor and helped him to lay out the path to the executive team. His projects increased in size to match the size of his dreams. He began managing power projects at Lawrence Livermore National Laboratories, the University of California at Davis, and Beale Air Force Base. He led power development for the sealed-environment research project near Phoenix, Arizona called the Biosphere II and then became chief engineer for the world-renowned, Superconducting Super Collider project. When that project was canceled by Congress, Bill developed a Texas-based power engineering office and became Texas area manager of PB Facilities Services. In 1996, he moved to area manager for the San Francisco and Oakland offices, with full responsibility for a technical staff serving more than 40 public-agency clients, including the Bay Area Rapid Transit District.

Well, his climb continued. But where many of us are looking for bridges to cross, Bill was looking for them to build. He found one in the East Span of the San Francisco-Oakland Bay Bridge that had been damaged by the earthquake. He built light rail systems, widened major highway exchanges, and extended the rapid transit system to the San Francisco Airport. So it is not wrong to say that Bill repaired burned bridges, built the high road, and extended the midnight train to the air to arrive at his present success.

So, let me jump to that present success that has brought us all here tonight. Bill Smith today is president of Parsons Brinkerhoff Quade & Douglass, the transport infrastructure arm of one of the oldest, largest, and most reputable engineering consulting companies in the United States. It is a company with over 3,000 employees handling 1,700 projects ranging in size from $50,000 to $14 billion, and earning $680 million in annual revenues. Before you meet Bill, let's hear from the Old Storyteller.

The Old Storyteller: Yadkin County is located in the foothills of North Carolina. The Blue Ridge Mountains serve as a distant backdrop for the gently rolling landscape that is Yadkin Valley. The county takes its name from the Yadkin River, which serves as its northern and eastern boundaries. Since it was founded in 1850, Yadkin has been composed of farming communities. While the population is growing, there are still many open spaces and woodlands. Old farmhouses still dot the countryside along with the new housing developments. Historical landmarks recall times gone by and the two interstate highways prophesy possibilities of the future. This is the fertile soil that nurtured Bill Smith.

He was born in the mid- forties, the oldest of nine children born to Mary Pauline and Sherwood Smith. He was a rather shy, asthmatic lad growing up, which probably accounts for his love of books. He is an avid but eccentric reader; he prefers math and science books to fiction. Says he, "I like books that make me think about the world we live in." We don't know for certain how Bill came to be who he is. All we can really tell you is that it is obvious that he was nurtured to success. He had extended family aplenty living in and around his neighborhood. And for three of his formative years, he was cultivated in the home of his doting grandparents. They put him in touch with the imperative of his roots.

With nine children growing up in a country home run by a no-nonsense matriarch and a harmonica-playing bluesman, your values are planted true and firm. Throw in an occasional heathkit for good measure, and give a growing boy access to some pliers and a vacuum tube radio set, and what he doesn't blow up is bound to teach him. This is the story of our Bill. So at the age of 13, he told his mother, "I'm going to be an electrical engineer!" She probably said, "That's great baby, after you clean up that mess you made." But she never, ever said he couldn't! Even at times when Bill's tinkering blew the house lights out, he was told to find a way not to do that, but never to stop his tinkering. And that is probably why and when he became so mesmerized by electrical power. It also might explain why Mrs. Anderson, his third grade teacher, kept telling him he had to focus. He never really did except in one area—the thought of becoming an electrical engineer was the only career on his radar screen.

At one time, Bill was the only Black student attending the R.J. Reynolds College Preparatory school. It was a time of race consciousness, but he was never affected by it. He believed as he was told: that race is like the air you breathe; it's always there, and so you inhale it and let it work for you, not against you. He yearned to attend Rensselaer Polytechnic Institute for his electrical engineering degree, which he knew he was going to get, but the money was low, so he settled on North Carolina A&T, at the time when it was growing students like Rev. Jesse Jackson Sr. and Ron McNair the astronaut. It was in this same fertile soil that Bill Smith was reared. Windsor Alexander, now the dean of Engineering at UNC, was a fellow student of Bill's at A&T and a math whiz. Windsor would frequently enchant fellow students, struggling over a complex formula, by working it out in his head in front of them. For the uninspired, this might have been intimidating and turned them away from their aim. But Bill is not made of such ordinary stuff. He took the enchantment as motivation and although he struggled through college, he graduated and now is proud to place that P.E. for registered professional engineer behind his name.

Bill Smith loves to give back that which he has earned and learned. For the earning, he is a model to be emulated. For the learning, the philosophy that brought him here is as relevant today and to us all as it was to him in his youth. He was told repeatedly that you are a gift to the world sent forth into the wilderness with all the tools you need to conquer it. If you should ever doubt it, remember this: there is no such thing as bad weather, just inappropriate clothing.

For these reasons, it is our delight to join with the Honorable Secretary Rodney Slater in presenting to you The Black Engineer of the Year for 2005, Mr. Bill Smith!

2005 Black Engineer of the Year Bill Smith: Thank you Rodney for your resolute support and for being here to share this moment with me and my family.

I want to thank my wife Sharon, my brother and sister, and all my friends and extended family who have been part of the support team, which has helped me to build this career.

I'm very honored to receive this award. I learned of its existence only recently, and am very gratified to have done so because it means that the word is getting out to the world that, like all other engineers, African Americans are in pursuit of excellence. That is the force that can break down any barrier to success.

Tonight's theme, "Seizing the Excellence Imperative," is very provocative. It reminds me of when I was growing up in North Carolina, I was fascinated by electricity. I was mesmerized by how all that power could be hidden. You couldn't see it, you couldn't smell it, you couldn't hear it or taste it—but you knew it was there, working in the background, because when the right contact was made, something was going to happen. That is about how I see the excellence imperative, a success potential waiting in the background ready to unleash its mighty power when the right connection is made.

My mother used to always tell us that we could never get ahead by being "as good as;" we had to strive to be "better than." Now, she was talking about performance, not status, and although I might have shrugged it off at the time—as teenagers are likely to do—her constant admonition became one of my imperatives, a silent but sure current running in the background until I made the right connections.

Another excellence imperative is to always work towards the greater good. I believe it is vital that each of us do some good in this world. There's a lot of room in the notion of "doing good" to cover what each of us can do every day to make a contribution. Doing good can include running a company, such as the one I am leading. In that context, I can do good by creating the environment wherein we deliver quality engineering solutions to our customers, provide professional opportunities for our employees, and make money for our shareholders. I particularly believe in the importance of caring for the people in our company—bringing them up, developing their technical excellence, letting loose the "true believers" who have great ideas for the company's future.

Doing good must extend into the community. I have always believed that community involvement is important for the individual and organization. Engineering excellence demands that we solve technical problems, sensitive to the fiscal and social concerns our clients encounter throughout a project's life, all while maintaining a strong focus on the community. It is imperative that we use our worldwide experiences to deliver local community solutions.

I also firmly believe, though, that it's important to have fun. Without a little fun, what's the point of it all?

I've enjoyed hearing the stories of all the wonderful accomplishments of the people recognized here tonight. I'm humbled to be honored with this award and truly appreciate all the people who've helped me along in my life. May we all go forward from this moment inspired to seize the excellence imperative.

I had the strangest dream last night that I was standing in this light on a stage before a throng of Black engineers.

And then this spirit in golden draping showed me a world quite ripe for taking. That very moment, I take control of all my fears.

Then she touched my heart with song, saying that I'd possessed all along the qualities that produce career longevities.

As she pointed to this Black engineer, she said, "He won the trophy of the year just for seeing excellence as an imperative."

Chapter 20

Linda Gooden
2006 Black Engineer of the Year

The following article, written by Roger Witherspoon, was first published in US Black Engineer & IT, *Conference Issue 2006.*

In 2005, the Federal Aviation Administration (FAA) needed a hand in a hurry.

There were more than 600,000 general aviation pilots crisscrossing the skies to thousands of commercial and private landing fields across North America, and monitoring them in increasingly crowded skies was difficult.

Pilots filed flight plans through 58 FAA call centers, which in turn provided weather data and other information, but these centers had limited capabilities and difficulty coordinating with each other. The FAA knew the system needed to be replaced.

Enter Linda Gooden, the founding president of Lockheed Martin Information Technology. She believed the FAA needed a fully integrated network of 20 centers, each carrying a redundant capability so that any one system could moniter a flight from origin to destination anywhere in the nation. The flight plans are automated, and if someone doesn't close out a flight plan—signaling that the plane arrived at its destination—the system alerts the nearest agencies.

It was not surprising that the FAA awarded Gooden's group the $1.9-billion contract. Under her leadership, Lockheed's IT division has become the world's leader in government information technology systems, with 14,000 employees in 70 domestic sites and 18 foreign countries, and more than $2.2 billion in revenues. What is surprising is that this company did not exist until 1994, when Gooden and four colleagues developed a business plan for a new venture within Lockheed.

Gooden knew from her research that the federal government intended to automate its massive infrastructure. A company with the right approach to automating federal office functions could grab a lot of future business.

Her initial request to Lockheed was small; she needed $200,000 in seed money and 57 employees to pursue outsourcing contracts with a target return of $11 million at the end of the first year. The company staked her new group to $600,000, and the new venture brought in $24 million in 12 months. It is the largest and fastest growing business unit in Lockheed's information sector.

Gooden's team has modernized the Social Security system, ensuring that the nation's elderly receive their benefits; digitized the FBI's fingerprint database, so millions of prints can be searched in minutes; automated the Navy's payroll system, so all personnel get paid, wherever they are in the world; developed the communications infrastructure for the 25,000 employees at the Pentagon; and eliminated mountains of paperwork by making the patent application process electronic.

"She is, first and foremost, one of the most driven executives I have ever encountered in my professional life," said Michael Camardo, Lockheed's executive vice president for Information and Technology Services.

Part of that drive comes from growing up on the south side of Youngstown, Ohio, in the midst of America's steel manufacturing heartland during a period of transition.

"It was a place," Gooden recalled, "where a person with a sixth grade education could get a good job with a decent wage and raise a middle class family.

"But when I was in high school in the 1970s, the steel mills started to shut down, and a lot of people lost their way of life. By the time I graduated, half the steel mills were gone, and a lot of people didn't have the opportunity to change their lives, and they lost their homes. It pointed out to me that people really needed an education so they would have a choice in life." She made her choice when she entered Youngstown State University and saw the installation of a new IBM 360 computer.

"When I saw that," said Gooden, "I decided that was what I wanted to do in life. I was mesmerized by the size of the machine. It took up six normal-sized rooms, though the computing power you have on your wrist today is probably greater than they had in those early machines. But it was clear to me that that was the direction the nation was moving in, and that's where I wanted to be. The heavy industries were going down, and this computer was offering a bright new future." Gooden received a degree in computer technology from Youngstown in 1977 and went to work writing software for General Dynamics. Three years later, Martin Marietta offered her a job developing the software for the Peacekeeper Missile. Then she was offered a chance to switch from military to business software, and develop and install a corporate-wide payment and personnel system in the company's Washington-area headquarters.

"I found it more challenging to deliver a human resource system than to deliver software that supported missile systems," Gooden said, "because you have to deal with people and personalities and likes and different understandings about the way things work. You have to get them to agree on a common set of requirements, and then develop a system that can be used by people with different needs and styles." That experience set the stage for the rest of her corporate life. Gooden began looking outside of the company for potential clients, and in 1988, she won a contract to modernize the Social Security Administration.

Social Security wanted to update its system in stages, which is more difficult than building a completely integrated system all at once. Gooden conceived of development architecture that was robust enough to allow for the integration of distinct modules as they were later developed.

"I take a lot of pride in saying that we build the systems that make America a better place to live," she said. "I was home for my grandmother's funeral, and both of my elderly aunts were there. Social Security checks are what they have to live on; that's their only form of income. We write the software that makes that happen. It's important to me when I do work for Social Security that I know I am doing something for my aunts." Gooden realized that there was a potential market for Lockheed's IT services throughout the federal government, and she held the planning session that launched Lockheed's information service. Three years later, Gooden sought and received permission to consolidate all information technology into her business area and create the extensive information services company that dominates government today.

Some of that work has been personally difficult. "We work in a safe industry, in nice offices," she said, "but 9/11 reframed our thinking about those people who are out there every day, working to support this nation.

"We had 400 people working in the Pentagon when the airplane hit the building. It took until about nine that evening to account for all of the employees. We had a team of people who walked from the Pentagon to our offices with the customers and worked all night so the Pentagon could be live the next morning. Those guys went into the Pentagon while it was burning and brought up those computers. You can't buy that kind of commitment." Gooden has also shown commitment to getting more young people interested in technology. "The digital divide is real," Gooden said, "and as the country becomes more and more diverse, the problem is not just for us as African Americans. It's a problem for the nation, as far as where the next generation of computer engineers and scientists is coming from." She is doing her part, working as the company's representative to both Morgan State and Hampton Universities, and she serves on the boards of the University of Maryland James Clark School of Engineering, the Prince George's Community College Foundation, the Maryland Business Roundtable for Education, and the Boy Scouts of America.

With Gooden as a guide, the youth who come her way could not have a better introduction into the world of tomorrow.

Printed in the United States
85703LV00005B/459/A